INCREDIBLE STORIES
FROM SPACE

A Behind-the-Scenes Look at the Missions
Changing Our View of the Cosmos

NANCY ATKINSON

editor of *Universe Today*

PAGE STREET
PUBLISHING CO.

TO RICK, ALWAYS

PAGE STREET
PUBLISHING CO.

First published in 2016 by
Page Street Publishing Co.
27 Congress Street, Suite 105
Salem, MA 01970
www.pagestreetpublishing.com

Distributed by Macmillan, sales in Canada by The Canadian Manda Group.

19 18 17 16   2 3 4 5

ISBN-13: 978-1-62414-317-5
ISBN-10: 1-62414-317-2

Library of Congress Control Number: 2016944631

Cover and book design by Page Street Publishing Co.
Cover image: Getty Images

Printed and bound in the United States

Page Street is proud to be a member of 1% for the Planet. Members donate one percent of their sales to one or more of the over 1,500 environmental and sustainability charities across the globe who participate in this program.

CONTENTS

RIDING ALONG ON THE JOURNEYS TO SPACE

On February 11, 2010, I stood alongside several scientists as we watched a rocket thunder off the launchpad at Kennedy Space Center (KSC) in Florida. Nestled inside the rocket was a robotic spacecraft these scientists had spent most of their careers working to make a reality.

For me, experiencing a launch up close and in person was something I had been anticipating for several months, although there was really no way of knowing ahead of time how indescribably exhilarating the launch experience would be. I highly recommend it for everyone.

Half a year earlier, I had begun making preparations to attend another launch at KSC, the STS-130 mission for space shuttle Endeavour to send six astronauts to install two new components on the International Space Station. Since NASA had announced in 2004 that the space shuttle program would be ending by 2011, I knew I wanted to be on hand for some of the final missions launched aboard this magnificent and storied spacecraft. After all, I was a journalist who specialized in space exploration and astronomy: what could be more poignant than to report on the end of an era in spaceflight?

By analyzing launch schedules, however, I soon realized I could maximize my trip and see additional launches by simply extending my stay. Plus, since I was living in Illinois at the time, the concept of spending a few weeks in Florida during the winter seemed like a really great idea.

My time on the Space Coast in 2010 ended up being a bucket list type of trip, where I not only had the opportunity to witness four different launches (two space shuttles and two robotic spacecraft), but also visit NASA facilities, get behind-the-scenes tours and insights and meet and interview dozens of astronauts, scientists, engineers and NASA officials. A highlight was the once-in-a-lifetime opportunity to stand directly on historic Launch Pad 39A at KSC, right underneath the massive space shuttle Discovery, poised to launch on its penultimate flight.

The Atlas V rocket carrying the Solar Dynamics Observatory begins rolling out to the launch pad at Cape Canaveral Air Force Station from a thirty story gantry called the Vertical Integration Facility. Credit: NASA

While human spaceflight is certainly compelling—and it has always been a big part of my reporting career—there is something about unmanned robotic spacecraft that has always tugged at my heart. These machines are our emissaries out into the cosmos, flung to faraway places that humans can't yet visit. I grew up hearing about spacecraft like Mariner, Viking and Voyager boldly going on some of the first-ever deep space missions and making monumental discoveries that changed our view of the solar system. They showed us worlds we previously could only dream about and artists could only imagine.

Now, every day we receive real, stunning images direct from our spacecraft stationed at other planets, dwarf planets, asteroids, comets and moons in our cosmic neighborhood. Whatever it takes to make it work, robotic spacecraft have done flybys, orbited, soft-landed, crashed into and roved about on various places in our solar system, all in the name of science and exploration. We've also launched big telescopes into space, putting them high above the murkiness of Earth's atmosphere to get a clear view of stars, distant galaxies and even other solar systems completely different from our own. These missions are showing us we live in an amazing, breathtaking universe.

Despite the fact that these are unmanned machines—made just of metal and circuitry—out there exploring the cosmos, there is still a human element to it all. Humans yearn to know the answers to questions about the heavens, and so we devise and build spacecraft. Human ingenuity calculates trajectories and ephemerides (computations of exactly where planetary bodies and spacecraft are at a given time) to set the spacecraft on course across interplanetary distances. The investigative human mind analyzes the data to make conclusions and discoveries. And even those of us humans who aren't rocket scientists are awed and amazed at the intrinsically beautiful images from space and wonder at marvelous new findings from faraway places. These machines allow us to explore the cosmos, all from the comfort of our home planet.

But getting a robotic spacecraft on a mission to space is no easy task.

The usual process is that astronomers and planetary scientists can toil for years in their field of study and somewhere along the line an unsolved question or deep mystery sparks an idea for an instrument, or even a full-up spacecraft with numerous instruments, to study the mystery. Preliminary mission concepts and schematics are drawn up with potential coinvestigators.

Then the scientists need to wait until a space agency like NASA (the National Aeronautics and Space Administration in the United States), ESA (European Space Agency), JAXA (Japan's space agency) or ISRO (the space agency in India) has what is called an **Announcement of Opportunity** (AO), where the space agency puts out a call for proposals for missions to space. Usually, however, the space agency is looking only for a certain kind of mission, so if the type of mission the scientists want doesn't fall under this specific AO (yes, space exploration is full of acronyms) they need to wait longer until the right opportunity comes along.

Finally, when the right prospect is available, the scientists assemble a team and write a proposal. Usually, scientists have to submit *several* proposals for their idea, and hope their mission makes the cut during several rounds of expert reviews. Eventually, a very lucky few can finally win funding from their space agency. Then happily—if not frantically—they need to arrange for the construction of the instruments and spacecraft, often with several revisions and modifications needed. A launch vehicle (rocket) needs to be arranged and secured. Oh, and then there's the chance the country's Congress or

Parliament decides it needs to cut their space agency's budget, and the mission that was thought to be approved gets put on hold or canceled outright.

The entire process can take years, even decades. And there are no guarantees that everything (the rocket, the instruments, the spacecraft itself) will work exactly as needed—indeed perfectly—in order for the mission to succeed.

But on that day back in February of 2010 came the perfect culmination of all the years of planning and work for the scientists. The spacecraft (the Solar Dynamics Observatory, which you'll read about later) launched successfully, a magnificent beginning to mission operations. Being able to witness the excitement and joy in the faces of the mission's science team is something I'll always treasure.

Over the years while writing for Universe Today, I've had the privilege of "riding along" on these missions to space: I've gotten to follow the missions closely, analyze the discoveries and meet some of the amazing people involved. I've also had the privilege of sharing with the public the incredible stories of the missions, as well as the stories of the scientists and engineers who make these missions go—those who conceive missions, who build the spacecraft, who study the data and who care for our robot emissaries heading out into the cosmos.

The tales of these machines and the humans behind them are quite compelling, filled with high drama and twists and turns along the way. As many space scientists have expressed, they never know what they'll find and they've learned to expect the unexpected.

This book provides a snapshot in time of stories about just a few robotic space missions in the early twenty-first century. Space missions begin and end, with some lasting for years and even decades; others just a few months. But as is the case in every human saga, the end of one story signals the beginning of another. The story of exploration is ongoing and connected.

Besides the ones detailed here, numerous other fantastic robotic space missions are ongoing, and I encourage you to find out more about them. And the final chapter in this book covers several new and upcoming missions.

Today, "riding along" on these missions to study the cosmos is available to almost anyone, not just journalists. Online access and connections through social media allow the general public to participate in events, ask NASA and other space agencies questions, view images and learn about discoveries, sometimes shared in real time as they happen.

Additionally, with the advent of citizen science projects, even the average person can do more than just follow along; anyone can now actually contribute to the scientific process. What could be more exciting than finding a hidden galaxy, a new supernova, previously unseen craters on the Moon or Mars or even a new planet orbiting another star? Those objects have already been found by average, everyday people and more discoveries await.

Come along for the ride . . .

Nancy Atkinson
Burnhamville Township, Minnesota
2016

CHAPTER 1

UNLOCKING PLUTO'S SECRETS: NEW HORIZONS

THEY SAID IT COULDN'T BE DONE

In typical Alan Stern style, he squeezed in time for a phone call during a drive between meetings while making a quick stop at home, too. A ratty Bluetooth connection left Stern's voice barely audible over background road noise and traffic, then car and house doors slamming, briefcases and drawers opening and closing. All the while he discussed his spacecraft's incredible, almost

Artist's concept of the New Horizons spacecraft during its encounter with Pluto and its moon, Charon. Credit: Johns Hopkins University Applied Physics Laboratory/Southwest Research Institute (JHUAPL/SwRI)

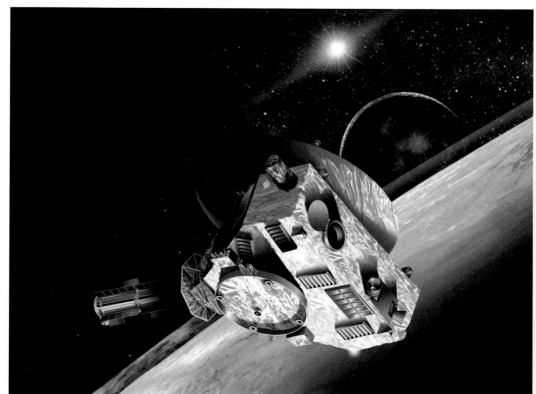

inconceivably successful mission to the far reaches of the solar system to study distant Pluto and its moons in July 2015.

"We did it," Stern said succinctly, never one to ramble on, talking about how he now responds to those who said the New Horizons spacecraft could never be built on time, on budget and still succeed. "We don't need to say anything more."

Stern, the principal investigator—and chief instigator—of the New Horizons mission, had been pushing for a mission to Pluto since the late 1980s.

"The first ten years we couldn't even get out of the starting blocks," he said. Other missions to Pluto had been proposed but were either canceled or not seriously considered. Frustrated, Stern started lobbying other scientists and even members of Congress, resulting in the National Academy of Sciences ranking a mission to Pluto as the highest priority for the first decade of the 2000s.

New Horizons' mission to Pluto was initially approved in 2001, but then In 2002 it looked as though NASA would have to scrap the mission for budgetary reasons. But space advocacy groups like The Planetary Society and others petitioned Congress to keep the mission alive, in part by using the power of letters from children. Seemingly, kids have always been enamored with the planet with the cartoon character name, even though Pluto is actually named after the Roman god of the underworld.

The plan worked and the mission to Pluto was reinstated. But still there were naysayers.

"When New Horizons was selected in 2001, people told us, 'you won but you lost,'" Stern recalled. "They said no one can build a mission at this price, one fifth the cost of the Voyager mission in the 1970s. No one has built an outer planet mission in this timeframe, just four years. Then, it's a ten-year journey to Pluto and you've only got one spacecraft, so there's a significant chance of failure. It just can't be reliably done."

But, Stern contends, the New Horizons team resolved to make it happen. "They pulled off something few people appreciate. It took dedication from 2,500 people around the country who worked all day plus nights and weekends for over fifteen years to make this dream come true. It's the stuff of history."

The dream was sending a spacecraft to explore Pluto and its moons. "Exploration always opens our eyes," Stern told me back in 2005, just months before New Horizons would launch. "No one expected it when previous missions found river valleys on Mars, or a volcano on Io, or lakes on Titan. What do I think we'll find at Pluto and Charon? I think we'll find something wonderful, and we expect to be surprised."

Stern's prediction—and his dream—came true.

NEW HORIZONS

New Horizons was the fastest spacecraft ever launched using a souped-up Atlas V rocket with extra boosters.

"We built the smallest spacecraft we could get away with that has all the things it needs: power, communication, computers, science equipment and redundancy of all systems, and put it on the biggest possible launch vehicle," said Stern. "That combination was ferocious in terms of the speed we reached in deep space."

The baby grand piano–size spacecraft sped away from the Earth at 36,000 miles per hour (about 58,000 km/hour), going nearly a million miles (1,609,344 km) a day in its 3-billion-mile (4.8-billion-km) journey to Pluto. But at that blazing speed, slowing down and going into orbit around Pluto would be impossible. Therefore, this first reconnaissance of the Pluto system would be a flyby mission, somewhat of a throwback to the early space missions like Mariner and Voyager that made the first flybys of Mars, Jupiter, Saturn, Uranus and Neptune in the 1960s, 70s and 80s.

New Horizons launched on January 16, 2006, and zoomed past the Moon's orbit in just nine hours. That same journey took the Apollo astronauts three days. Then New Horizons made the 500-million-mile (800-million-km) journey to Jupiter in just 13 months—faster than any of the seven previous Jupiter-bound missions—as the spacecraft swung past the giant planet for scientific studies and an all-important gravity assist in February 2007, boosting New Horizons' speed to 52,000 mph (83,600 km/h).

Despite its speed, it took nearly nine-and-a-half years for New Horizons to cross the expanse of the solar system and reach the region around Pluto. The true mission would start about five months out, before what's called **closest approach**, so in total there would be a six-month-long reconnaissance as New Horizons approached, flew by and then looked back at the Pluto system, with closest approach on July 14, 2015.

New Horizons launched on January 16, 2006 from Cape Canaveral Air Force Station, Florida on an Atlas V 551 rocket. Credit: Scott Andrews/NASA

The New Horizons science payload consists of seven instruments designed to investigate the global geology, surface composition and temperature, and the atmosphere of Pluto and its moons. Credit: NASA/JHUAPL

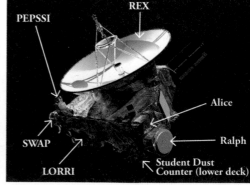

Since New Horizons' journey takes it so far from the Sun, solar panels wouldn't be feasible to provide enough electrical power. Instead, a **radioisotope thermoelectric generator (RTG)** produces heat from the natural decay of non-weapons-grade plutonium-238, and the heat is converted into electricity. The heat also keeps electronics and other parts of the spacecraft from freezing in the cold depths of space.

After Pluto, New Horizons continues on to explore the Kuiper Belt, a region out past Neptune, about 30 to 50 times Earth's distance from the Sun, or 2.5 to 4.5 billion miles (4.5 to 7.4 billion km) away, where Pluto is located. This belt is similar to the asteroid belt found between Mars and Jupiter; however, some objects in the Kuiper Belt tend to be more icy rather than rocky. This is where icy short-period comets (such as Halley's Comet) originate.

The spacecraft carries a robust payload of imagers, spectrometers and other scientific instruments to map the surface of Pluto and its moon Charon, study their composition and analyze any atmosphere around either body.

At launch, New Horizons was billed as the "first mission to the last planet." But that designation changed just seven months later.

PLANET OR NOT?

Several things changed, actually, while New Horizons was in flight.

When the spacecraft launched, Pluto was classified as a planet. But ever since its discovery in 1930 by astronomer Clyde Tombaugh, the distant little Pluto was always considered a bit of an oddball: it didn't fit in with the rocky terrestrial worlds of the inner solar system, and the tiny, icy, 1,473-mile (2,370-km)-wide world definitely didn't match up with the gas giants of the outer solar system. Pluto's orbit was unique, what astronomers call an **eccentric orbit** due to its highly elongated and angled path around the Sun compared to the relatively circular and flat orbits of the other planets.

Pluto's eccentric orbit compared with the rest of the solar system. Credit: NASA

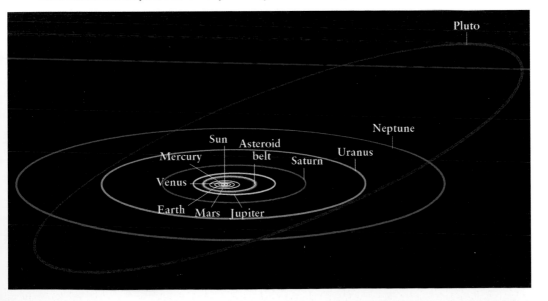

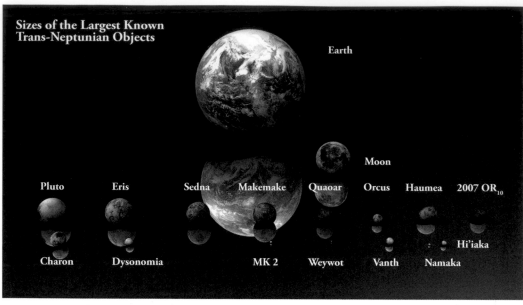

Size comparision of the largest known trans-Neptunian objects. Credit: Kevin Gill. Earth: NOAA/NASA/Univ. of Wisconsin, Suomi NPP. Moon: NASA. Pluto and Charon: NASA/JPL/SWRI.

Of course, Pluto was so far away even the best pictures available taken by the Hubble Space Telescope were just pixelated blobs. No one really knew what Pluto was like. One of New Horizons' goals was to understand how Pluto and its moon Charon fit in with everything else.

Decades before the mission, Stern and other planetary scientists theorized that Pluto wasn't alone out there in the far reaches of our cosmic neighborhood. It didn't make sense that the solar system would abruptly run out of material after Pluto. Plus, since the Kuiper Belt was thought to be the birthplace of comets, why couldn't other objects be there, too?

In a paper published in 1991, Stern proposed there could be hundreds of smaller yet-unseen icy bodies like Pluto and Triton—Neptune's moon and a virtual twin of Pluto, and thought to be a Kuiper Belt Object (KBO) that wandered out of the belt, captured by Neptune's gravity. Stern used the term **dwarf planet** in his paper to indicate a new subclass of planets that might be needed to properly classify Pluto and the large KBOs they expected to find.

Just a year later in 1992 came confirmation with the first true discovery of a KBO, a tiny object about 100 miles (161 km) across. Since then, astronomers have found hundreds of KBOs, and scientists now estimate there could be thousands of such bodies. This put Pluto into context, as it seemed to be just one of a much larger class of tiny, icy bodies. And because KBOs are thought to be unaltered since the birth of the solar system 4.5 billion years ago, studying Pluto would provide clues to the conditions that prompted planets to form. That led to Stern's push for a mission to Pluto.

A decade later, with the advent of better telescopes and other improved observation techniques, scientists found bigger objects more like Pluto. Astronomers Mike Brown and Chad Trujillo from Caltech discovered an object about half the size of Pluto named Quaoar (pronounced kwa-war, named for another ancient god), followed in quick succession by Sedna in 2003, Haumea in 2004, and Eris and Makemake in 2005. Eris was the clincher, as it was about the same size as Pluto.

The debate that had been simmering for years now heated up in earnest: should all these large KBOs be classified as planets, too? Or would it be better to instigate a new class of planets that encompassed Pluto and it's nearer, more similar companions?

Curiously, at that time there wasn't an official definition of a planet. But with the new KBO discoveries, it seemed time to create one, as well as decide what to do with Pluto and all these new objects. The International Astronomical Union (IAU) is usually the arbiter and decision-making body in cases like this, so in August of 2006 at an IAU Assembly meeting, it was put to a (controversial) vote.

There were three choices:
1. Add Eris, Makemake and the largest asteroid Ceres to the planet club, bringing the total number of planets in our solar system to twelve.

2. Keep the total at the familiar nine, and not really address the new discoveries.

3. Drop the number of planets down to eight, kicking Pluto out and creating a new class of objects called dwarf planets.

The difference—and the kicker—between Stern's use of that term in 1991 and the IAU's word choice in 2006 was that the IAU said *dwarf planet* wasn't just a subclass of planets; it wasn't a planet at all. The IAU's definition inferred Pluto and its pals were a completely different type of object.

To help make the decision, the IAU also created the definition of a planet, with an object needing to meet three requirements:
1. It is in orbit around the Sun.

2. It has enough mass and gravity to pull itself into a spherical shape.

3. It has cleared the neighborhood around its orbit.

While Pluto meets two of these criteria, it fails on the third. **Clearing the neighborhood** means the planet has become gravitationally dominant; there are no other bodies of comparable size other than its own moons in its vicinity in space. Since Pluto shares its orbital neighborhood with other KBOs, that demotes Pluto to the newly created classification of dwarf planet.

This set off a war of words between astronomers and planetary scientists, with Stern at the forefront of the fight for Pluto to retain its planetary status.

"Astronomers aren't experts in planetary science, and they basically passed a bunch of BS off on the public back in 2006 with a planet classification so flawed that it rules the Earth out as a planet, too," Stern said. "A week later, hundreds of planetary scientists, more people than at the IAU vote, signed a petition that rejects the new definition. If you go to planetary science meetings and hear technical talks on Pluto, you will hear experts calling it a planet every day."

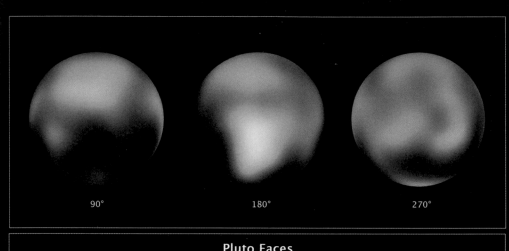

Prior to the New Horizons mission, these were the best and most detailed views of Pluto's surface, as constructed from multiple NASA Hubble Space Telescope photographs taken from 2002 to 2003. Pluto is so small and distant that the task of resolving the surface is as challenging as trying to see the markings on a soccer ball 40 miles (64 km) away. Credit: NASA, ESA and Marc Buie (Southwest Research Institute)

Stern, along with other scientists who had been studying Pluto for most of their careers, took the reclassification as a direct insult. To this day, this group of scientists—sometimes called the **Pluto Mafia**—as well as many other scientists, still reject the IAU's definition of a planet both in terms of defining a dwarf planet as something other than a type of planet and in using orbital characteristics rather than surface features and other fundamental properties to define a planet.

But while Pluto was demoted in scientific terms, the public seemed to embrace Pluto even more. Lovers of Pluto as a planet protested and aired complaints. Stern and the Pluto Mafia held their ground, hoping that New Horizons might one day change the classification defining their favorite little world.

The other change that occurred during New Horizons' flight was the discovery of four more moons orbiting Pluto. Charon, the largest, had been discovered in 1978 and is almost half the size of Pluto. The two bodies actually orbit each other around a common center of gravity, and therefore Pluto and Charon were sometimes referred to as a binary planet (today they are sometimes referred to as a double dwarf planet system).

In 2005, as New Horizons scientists used the Hubble Space Telescope to search the region around Pluto to help prepare for the mission, they discovered two new tiny moons at Pluto, now named Nix and Hydra. Then in 2011, Kerberos was found between the orbits of Nix and Hydra, and in 2012, another little moon named Styx was found.

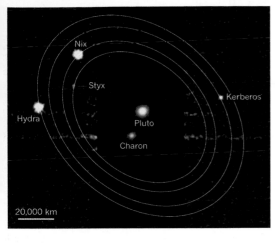

Nix

Styx

Kerberos

Hydra

Pluto

Charon

20,000 km

This image taken with the Hubble Space Telescope shows Pluto's smaller moons Styx, Nix, Kerberos and Hydra, which orbit the gravitational center between Pluto and Charon. Credit: NASA/STScI/Mark Showalter

The discovery of all these moons made the New Horizons team realize there could potentially be more moons and possibly a ring or debris field generated by a theoretical impact in Pluto's busy system billions of years ago. Any debris would pose a hazard for the spacecraft, and so the team began a thorough search using Hubble and later, New Horizons' instruments. Also, they needed a plan for mitigating any potential strikes by debris, so that task was put on the already long to-do list as New Horizons moved steadily towards Pluto.

BUSY

Hal Weaver is the project scientist for New Horizons and Stern's longtime colleague and friend. I visited him in his office in early 2016 at the Applied Physics Laboratory at Johns Hopkins University in Laurel, Maryland—where New Horizons was built and where the spacecraft's Mission Operation Center is located.

Weaver mused how psychological testing has shown that while everyone thinks they can multitask, most people aren't good at it.

"Alan is the exception," he said. "He's the one person I know who is able to handle many, multiple projects at once."

In addition to starting and overseeing New Horizons, Stern is an associate vice president at the Southwest Research Institute (SwRI) in Boulder, Colorado, an organization that does research in several areas, including planetary science and astrophysics. He also was the principal investigator on an instrument on the Lunar Reconnaissance Orbiter (which you'll read about later on page 180) called LAMP, and another called Alice on ESA's Rosetta mission to a comet. He's a consultant to various universities and aerospace firms. In 2007 and 2008, he directed all of NASA's space and earth science programs as a NASA associate administrator. Additionally, he cofounded a company called World View to take tourists to the edge of space in high-altitude balloons, and another company, Golden Spike, which helps coordinate commercial missions to the Moon. He also helped start Uwingu, an organization that raises money for astronomy research by unofficially and somewhat controversially selling the naming rights to exoplanets and craters on Mars. He also is a proponent of suborbital research and has helped design instruments for studies on board future flights of private companies like Virgin Galactic. Stern trained as an astronaut in the 1990s but never had the chance to fly, so he hopes to be on board several suborbital science flights.

Because of all this and more, Stern was listed among *TIME* magazine's 100 most influential people in the world in 2007 and is a candidate again for the same honor in 2016.

Alan Stern, New Horizons principal investigator speaking at a press conference at NASA Headquarters. Credit: NASA/Joel Kowsky

New Horizons Project Scientist Hal Weaver, speaking at a media briefing. Credit: NASA/Bill Ingalls

"I'm not sure how Alan does it all," Weaver said. "I think his administrative assistant has an administrative assistant! But he's an amazing guy. Period. You can't find anyone who is more energetic than he is. He is involved in so many things I can't keep up with it . . . or him."

Weaver himself feels like he hasn't had a spare minute since joining the New Horizons team in 2002. There was a rush to make sure New Horizons launched by 2006 to exploit an advantageous alignment between Jupiter and Pluto. The gravity assist from Jupiter would shave three years off the travel time to the outer solar system.

"There was a lot of incentive to get the spacecraft ready to launch on time," Weaver recalled, "but that required meetings almost every day throughout the development period. It was unbelievable. I just remember so many Sunday morning meetings as we were putting all the instruments together and integrating them on the spacecraft."

Another confirmed workaholic, Bill Gibson from SwRI was the payload manager, and he notoriously scheduled Sunday morning meetings. No problem for Weaver and Stern, at least.

"Alan and I are early birds so we were always up. And we love weekends because that just gives us time to do more work," Weaver laughed.

Even after launch, the pace didn't slow, which might seem surprising with the nine-and-a-half-year flight time to Pluto.

"We had been so busy getting the hardware together and to the launchpad," Weaver said, "we still had to plan in detail all the observation sequences at Pluto. Plus we had to get ready for the Jupiter-observing campaign, and we'd arrive there in just thirteen months."

Not only did Jupiter provide a gravity assist, it also gave the team an opportunity for a dry run for the Pluto encounter.

"There wasn't a lot of pressure for observing Jupiter," Weaver said, "because we weren't expecting any great science results to come out of it."

A montage of New Horizons images of Jupiter and its moon Io, taken during the spacecraft's Jupiter flyby in early 2007. Visible on Jupiter is the planet's Great Red Spot. On Io, a major eruption is in progress at the northern volcano called Tvashtar. This image appeared on the cover of the Oct. 12, 2007 issue of Science *magazine. Credit: NASA/JHUAPL/SwRI*

But the team—used to pushing the pedal to the metal—piled on the observations at Jupiter.

Weaver is also the lead scientist for the LORRI instrument on New Horizons, the Long Range Reconnaissance Imager, which has a telescopic camera to take global images from far away and provide close-up images for geological data when nearer to the target.

"We built it to have the sensitivity to allow us to observe the Pluto system," explained Weaver, "but sunlight there is 1,000 times fainter than at Earth. So LORRI is especially sensitive, and we thought it might be too sensitive to do imaging at Jupiter."

But in tests shortly after launch, Weaver and his team discovered just how versatile an instrument they had built. All they had to do was cut exposure times to very small values they initially didn't think were feasible.

"Sure enough, it worked beautifully and we jam-packed that flyby," Weaver recalled. "We got all kinds of great science at Jupiter, and we made the cover of *Science* magazine [analogous to rock stars making the cover of *Rolling Stone*] just for serendipitous observations because we happened to be flying past Jupiter."

With that success, the team pushed for going beyond the original plans for the type and amount of observations at Pluto.

"Our project manager Glen Fountain kept telling the scientists we just wanted to pile on more work for the engineers," Weaver said smiling. "But since this was our only opportunity to get the science at Pluto, we just kept pushing to do more."

The science team wanted to double the amount of observations that were originally planned at Pluto. But the entire observation sequence for all the instruments needed to be meticulously planned out to the second, so New Horizons' flight time was filled with coordinating and testing the plans.

Additionally, they had to deal with the potential debris problem.

"If we hit a dust-size one millimeter particle, it would blow a hole in the spacecraft and we could potentially lose the mission," Weaver said. "But as we got deeper into that particular problem, we felt better and better about it."

This illustration shows some of the final images used to determine that the coast was clear for New Horizons' flight through the Pluto system, taken with the Long Range Reconnaissance Imager (LORRI), on June 26, 2015, from a range of 13 million miles (21.5 million km) to Pluto. Credit: NASA/JHUAPL/SwRI

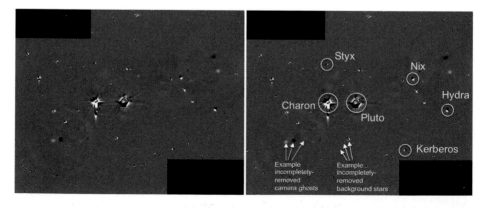

After analyzing the spacecraft, they realized the likelihood of mission failure from a particle hit was lower than originally thought. They also designed a new, safer flight path and, if needed, they could fly the spacecraft with its communications dish in front, serving as a deflector shield.

"We got the probability of mission failure due to particle hits down to less than a tenth of 1 percent," said Weaver, "so we felt good, but we still had to exercise appropriate caution. We kept searching and by the time we finally reached Pluto, we had already searched the Pluto system one hundred times deeper with LORRI than what Hubble could have done, and we still didn't find anything."

Another challenge came in navigating the spacecraft because scientists didn't know exactly where Pluto would be when New Horizons arrived. Pluto takes 248 years to complete an orbit around the Sun, and since astronomers have only known about Pluto's existence since 1930, they've only seen a small portion of its orbital path.

"This is a challenge not all missions have to deal with," said Michael Buckley, public information officer for the New Horizons mission. "Usually you know exactly where your target is. We had to be so precise in the timing and accuracy of the flyby point, coming within an imaginary rectangle in space measuring just 60 by 90 miles (100 by 150 km). And you had to deal with a delay in communications, since a radio signal—even traveling at light speed—requires 4 ½ hours to travel from New Horizons to Earth."

Navigators for New Horizons would be constantly refining their calculations throughout the mission, until just days before the flyby. They had to get it right or the spacecraft's observations might miss their target.

While the entire team was busy, the spacecraft itself spent about two thirds of its flight time asleep, in hibernation. This saved wear and tear on spacecraft components and reduced the risk of system failures at the critical Pluto flyby. New Horizons was awakened at least once a year for systems checkout and instrument calibrations.

LOST CONTACT

Alice Bowman is the MOM of the MOC. Translated from NASA acronym-speak, she works as the mission operations manager at the Mission Operation Center for New Horizons, located at the Applied Physics Lab.

Except for the whir of computer fans, the MOC was quiet on the day she gave me a tour of the facility in Building 13 at APL, and only display screens lit the room. Quite different, Bowman said, from what it was like on July 14, 2015—Pluto encounter day—when the MOC and an adjoining room were "bursting at the seams with scientists, cameras and NASA dignitaries."

But now the quiet of the control room exemplifies one reason why NASA approved this mission: "Because we were able to lower the costs to something that fit into their budget," Bowman said. "Hence, no people here at the moment."

New Horizons Mission Operations Manager Alice Bowman at a media briefing following New Horizon's closest approach to Pluto. Credit: NASA/Bill Ingalls

Additional cost savings came from sharing the command space with the MESSENGER mission, which ended in April 2015 after the spacecraft orbited Mercury for four years.

"When the MESSENGER team complained they had a twenty-minute round-trip delay in communications, we just looked at them and said, 'Really?'" Bowman said with a smile.

As mentioned, at the time of the Pluto flyby, two-way communication between New Horizons and Earth required a 9-hour round trip—4 ½ hours to the spacecraft and another 4 ½ back. Since radio signals travel at light speed (186,000 miles per second [300,000 km per second]), this exemplifies Pluto's great distance from Earth, nearly three billion miles (4.8 billion km). And now as New Horizons continues to zoom farther and farther away, the communications delay grows longer.

"Working with a spacecraft so far away is a challenge," Bowman said. "I always say you need to have a split personality when you work in ops [mission operations] because of all the variations in time. When you send a real-time command from Earth, you have to know where the spacecraft will be in the future."

Luckily, New Horizons has been a relatively trouble-free spacecraft.

Except for one day . . .

On July 4, 2015, just ten days before the Pluto flyby, Bowman and her team arrived early at the MOC so they could upload very special instructions to New Horizons. Called the **core load,** this sequence of commands covered the intense period of observations for ten days: seven days on approach, the big flyby day observation sequence and two additional days of observations as New Horizons looked back at the Pluto system. This would entail a total of 20,799 commands, including 461 scientific observations, minor trajectory corrections and scans for hazardous debris.

"We started loading the commands at 4:30 in the morning," Bowman said, "and with the round-trip light time, we could see the commands being received by the spacecraft about nine hours later."

All of a sudden, they lost communications from New Horizons.

Alice Bowman and Karl Whittenburg, New Horizons chief mission operations planning engineer, watch for incoming data from the Pluto-bound spacecraft. Credit: NASA/JHUAPL/SwRI

Bowman said her first thought was something was wrong with the receiving station on Earth. All planetary missions communicate through the **Deep Space Network (DSN)**, a system of extremely sensitive deep space communications antennas at three locations: Goldstone, California; Madrid, Spain; and Canberra, Australia. Their strategic placement approximately 120 degrees apart on Earth's surface allows for constant observation and communication with spacecraft as Earth rotates.

"We had experienced anomalies previously in the mission," Bowman said, "and normally the spacecraft is fine, but there's just a glitch in the ground station communications."

But a thorough checkout revealed the ground systems were fine. There was something wrong with New Horizons.

"I had a horrible feeling in the pit of my stomach," Bowman recalled. "But then you know you've got a job to do, and so you take a deep breath and say, 'OK, let's put into practice all those things we've trained for and learned from the past nine-and-a-half years.' And I knew we had an amazing team, so I just focused on the positives."

Stern, Weaver and other mission managers were called in and Bowman had already assembled her team of experts: Karl Whittenburg, the chief mission operations planning engineer; Chris Hersman, the mission systems engineer; Gabe Rogers, the guidance and control engineer; Steve Williams, the software lead for the onboard flight computer; and Brian Bauer, the lead for New Horizons' autonomy software.

Most critical was to re-establish communications with the spacecraft so they could access information on the health of the spacecraft and determine if the command sequence for the Pluto flyby was successfully uploaded to New Horizons. The flyby was going to start in three days, so there wasn't much time.

All spacecraft are programmed with a special operations system called **safe mode**. This is engaged when the spacecraft's autonomy system detects a severe problem requiring intervention from mission operations. In some instances, the autonomy system will cause a switch from the main to the backup computer and the spacecraft is programmed to point its communications antenna at Earth and transmit telemetry on its status and that it needs help.

This is what happened on New Horizons. Within 77 minutes, communications were re-established with New Horizons and the MOC started to again receive telemetry.

"I don't ever want to relive that 77 minutes of no communications," Bowman said, "but it was a wonderful feeling to regain the signal from the spacecraft. We knew then we could fix the problem but the question was, could we fix it in time before the planned start of the flyby sequence?"

After analyzing the telemetry, the team realized the computer was trying to do several things at the same time, which was more than the processor could handle. The spacecraft's computer was compressing data, loading software and also downlinking data to Earth.

"These were very intense operations commanded at the same time and the computer processor could not perform those operations fast enough, " Bowman said. "The autonomy system detected that the computer was falling behind and decided that there was something wrong with the main computer. It then commanded a switch to the backup (or secondary) computer. We lost the downlink from the primary side of the spacecraft because it had switched to the secondary side. Once we were able to look at the data, we figured out what was happening and put together a recovery plan."

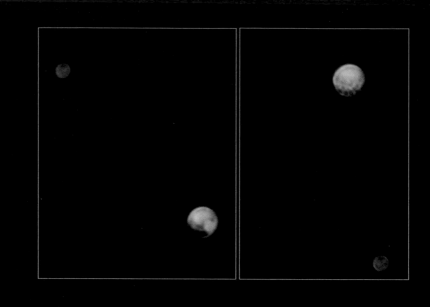

Getting Closer. Pluto shows two remarkably different sides in these color images of the planet and its largest moon, Charon, taken by New Horizons on June 25 and June 27, 2015. Credit: NASA/ JHUAPL/SwRI

It took nearly three days of painstaking work, but in the end, the core load of commands successfully uploaded to the primary computer. The observation sequences for the Pluto flyby would take place as planned.

"Even though it was an extremely tense time," Bowman said, "it ended up being a very gratifying experience with the team because we worked through the problem."

Weaver said the anomaly "made us all a little more circumspect, wondering if something else was out there lurking, waiting to happen."

One lurking question was how precisely the navigational calculations had brought the spacecraft within the desired target area. Bowman said the challenge of operating under that environment—and being able to respond to what the navigators were finding and perform correction maneuvers—was daunting.

"You don't want to correct the trajectory when you are too close to your target—Pluto— because you don't have time for that trajectory to 'settle' and to figure out exactly where you are," she said. "Our last opportunity for the correction maneuver was going to be July 4. After doing maneuvers, the only thing you can do to influence the observations is to adjust the timing of when they occur. That was part of what we were uploading when we lost communications."

But when the first optical navigation images from LORRI were beamed to Earth after the recovery, they showed New Horizons was still directly on course for Pluto. And Pluto—although still blurry in the images—was starting to look very interesting.

THE STORY OF THE ICONIC "HEART" OF PLUTO

On July 13, just before midnight—the night before closest approach—Weaver sat in his office, bleary-eyed, continually clicking on an icon on his computer to refresh it. He was waiting to receive a very special set of images, the final full global views of Pluto taken with LORRI before the spacecraft was too close for the entire planet to fit in the camera's field of view.

"We wanted to present this image to the science team and to the world in the morning," Weaver said, "and we had put together an expert image processing team who was going work through the night to get it ready for release. I was in charge of getting the data to that team, but I was going to wait and see the image together with the rest of the science team at a special meeting we had scheduled for 6:00 a.m. the next morning."

Weaver recalled how everyone had been working "like crazy" the entire week, definitely sleep deprived. This image would arrive from the spacecraft late in the day after an extremely busy time preparing for the momentous Pluto flyby.

"I felt like an idiot as I kept pinging my computer, looking at a certain directory to see if the images had arrived, but they hadn't."

After calling another member of the team, Weaver, exhausted, realized he had been looking in the wrong directory for about 45 minutes.

He slapped his forehead, and finally found the file. True to his word, Weaver didn't look at it—he put it on a jump drive and walked down the hall to the Imaging team. He decided to let them do their work, and went back to his office. A few minutes later, the imaging team walked in looking confused. Weaver had given them the wrong file, an earlier picture of Charon.

"I said, 'OK, let's look at this image together to make sure I get it right,'" Weaver said. The team gathered around Weaver's computer screen, the first time human eyes would see Pluto this close, at such a high resolution.

"We just gasped, our jaws hit the floor," Weaver said. "I ended up dropping the F-bomb and it wound up in print—I felt bad about it later—but we were absolutely amazed! It was just a raw image, not colorized, but we saw this huge, beautiful heart-shaped smooth area in the middle of Pluto! And all around the periphery you could see so much detail indicating that, wow, this was

so complex, so rich. We knew that we had a spectacular result on our hands."

New Horizons had just transformed Pluto from a pixelated blob—as seen by the best telescope ever built—to a spectacular world full of diversity and complexity. Plus, it seemed that Pluto was sending us its love.

This image of Pluto from the Long Range Reconnaissance Imager (LORRI) aboard New Horizons was taken on July 13, 2015, when the spacecraft was 476,000 miles (768,000 km) from the surface. This is the last and most detailed image sent to Earth before the spacecraft's closest approach to Pluto. Credit: NASA/ JHUAPL/SwRI

Members of the New Horizons team react to seeing the spacecraft's images before closest approach later in the day, Tuesday, July 14, 2015 at the Johns Hopkins University Applied Physics Laboratory (APL) in Laurel, Maryland. Credit: NASA/Bill Ingalls

FLYBY DAY

Downstairs from Weaver's office, the New Horizons science team gathered in an auditorium at 6:00 a.m. on July 14, 2015.

"We brought the new images of Pluto up on the big screen and the reaction from everyone was so cool," Weaver said. "These are the scientists that have been waiting so long for this. Pluto looked amazing, and these images were 1,000 times better than the best images we had from the Hubble Space Telescope."

Later, the science team was joined by their families for a special celebration as New Horizons made its closest approach at 7:49 a.m., coming within 7,700 miles (12,500 km) of Pluto's surface.

But whatever New Horizons was doing at that moment was unknown by the mission operations team. Communications with the spacecraft were deliberately suspended so the spacecraft could focus solely on gathering data. All they could do was hope the preprogrammed observation sequences were ticking off in order. The team wouldn't hear from New Horizons until later that night, when the spacecraft was already well past the Pluto system.

It was a long, tense 22-hour waiting period, but finally, precisely on schedule at 8:53 p.m., New Horizons phoned home telling the mission team—as well as people around the world listening in—that all went well. The message was a 15-minute series of engineering status messages—no science data—beamed back through the DSN.

New Horizons team members and guests count down to the spacecraft's closest approach to Pluto on July 14, 2015, at the Johns Hopkins University Applied Physics Laboratory in Laurel, Maryland. Credit: NASA/Bill Ingalls

As Weaver listened to each of the spacecraft subsystem leads reporting their data, he was waiting for two specific reports.

The first was from Brian Bauer, in charge of the New Horizons' autopilot software for the special encounter mode for the flyby. "When he said, 'no flags set,' that meant as far as the computer was concerned, everything was great and nothing anomalous had occurred during the flyby," Weaver said. "That was great to hear because there were so many things that could have gone wrong."

The second was from Steve Williams, the command and data-handling lead. "He indicated that from the telemetry, he could tell that all the data gathered during the flyby had been written to the solid-state recorders," Weaver said. "So that meant that not only had all observations successfully executed, but we also had all data on the spacecraft."

Only then did Weaver relax and breathe a sigh of relief, knowing New Horizons had crossed the equatorial plane of the Pluto system where the most dangerous particle impacts could have occurred.

Remarkably, despite all the unknowns in Pluto's exact location, after traveling nine-and-a-half years and 3 billion miles (4.8 billion km), the spacecraft came in just 80 seconds early and 50 miles (80 km) low from original expectations. The navigation team had hit the target 'box' and the correct compensations for the observations had uploaded successfully to the spacecraft.

Bowman then looked at the rest of the data coming in and announced, "We have a healthy spacecraft, and we're now outbound from Pluto."

New Horizons' flyby of Pluto occurred 50 years to the day from the very first planetary flyby, the Mariner 4 mission that flew past Mars in 1966. New Horizons was the farthest any spacecraft had been from Earth to conduct its primary mission.

New Horizons Flight Controllers celebrate (top) after they received confirmation from the spacecraft that it had successfully completed the flyby of Pluto, and bottom, Alan Stern triumphantly joins the team on July 14, 2015, in the Mission Operations Center (MOC) Credit: NASA/Bill Ingalls

"This is the lesson we've learned from 50 years of space exploration," Weaver reflected. "You have to go there if you really want to take that giant leap compared to what you can do by just observing from Earth. You have to send a spacecraft out there to these bodies."

THE AFTERMATH

The science data wouldn't start trickling in until the following day, and over the next few weeks only about 5 percent of the data would arrive, just low resolution images at first. It wouldn't be until September 5, 2015, that New Horizons began an intensive downlink session that would last over a year, sending back the tens of gigabits of data the spacecraft collected and stored on its two digital recorders during the flyby.

"This is what we came for—these images, spectra and other data types that are going to help us understand the origin and the evolution of the Pluto system for the first time," said Stern. "We're seeing that Pluto is a scientific wonderland. The images so far have been just magical. It's breathtaking."

A color version of the full image of Pluto taken by LORRI on July 13, 2015. Credit: NASA/ JHUAPL/SwRI

Charon in enhanced color. New Horizons captured this high-resolution enhanced color view of Charon just before closest approach on July 14, 2015, taken with the Ralph/Multispectral Visual Imaging Camera (MVIC); the colors are processed to best highlight the variation of surface properties. Credit: NASA/ JHUAPL/SwRI

Only months after the flyby did the team have the time to fully reflect on the Pluto encounter.

"I remember Alan coming into my office at the end of July," Weaver said. "He asked me if it was sinking in yet. We just hugged each other, and we said, 'My gosh, we really did it!' It's amazing it all went so flawlessly. This really was a once-in-a-lifetime opportunity and it has been a spectacular, wonderful journey."

For Stern, the frantic pace hasn't let up. Between meetings, presentations and traveling, he squeezes in time for studying the data from New Horizons.

"The entire day of the Pluto encounter was very real and surreal at the same time," he said. "I kept saying, 'I can't believe we are finally doing this, we've been talking about this for so long.' Honestly, sometimes I still pinch myself."

The best part was sharing the experience with his team of scientists and engineers who had worked long and hard for this outcome.

Glen Fountain, center, New Horizons project manager, walking to a press conference with Jim Green, NASA Planetary Science Division Director, left, and Alice Bowman, right, after the team received confirmation from the spacecraft that it completed the flyby of Pluto. Credit: NASA/Joel Kowsky

"This is a big team of very dedicated people," Stern said, "and after fifteen years together, we grew to be a family, in many respects. And we did it, it was just spectacular. Of course Pluto turned out to be beyond our wildest expectations, so not only was it a technical achievement—with the public really engaged in the mission—but now scientifically it's just been incredible! We've been blown away."

For Bowman, the entire day of the flyby is a blur. She and her team had been running on little sleep and the long day of the encounter was intense.

Her favorite memory of that day—the one she keeps coming back to—was a special moment she shared with her mission operations team after they had received the communications from the spacecraft and knew the flyby had succeeded.

"After all the dignitaries had left the MOC," Bowman said, "just the team was here and we shared a toast together. Then we made our way down to the Kossi Center [the Kossiakoff Conference Center at APL] for the press conference. It's a good five-minute walk, and even though we were a big group, in a way it was a private moment because we weren't being interviewed or at a press conference. It was a time we could look at each other and say, 'Wow, we did this together.'"

PLUTO UP CLOSE

Even though New Horizons is now well past Pluto, continuing on a beeline into the Kuiper Belt, the postflyby period remains a critical and busy time for the mission science team.

"We're now getting all the science we went there for," Weaver said, "so we've had no rest. We continue to have our noses to the grindstone."

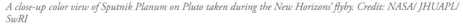

A close-up color view of Sputnik Planum on Pluto taken during the New Horizons' flyby. Credit: NASA/ JHUAPL/ SwRI

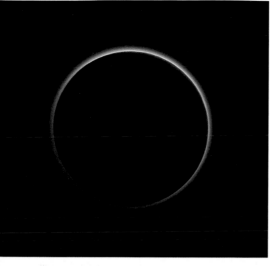

This image was taken just 15 minutes after New Horizons' closest approach to Pluto on July 14, 2015, as the spacecraft looked back at Pluto toward the Sun. The backlighting shows the deep haze layers of Pluto's atmosphere extending all the way around Pluto. Credit: NASA/ JHUAPL/SwRI

Because of New Horizons' distance and low power output (the spacecraft runs on just 200 watts of electricity), it has a relatively low downlink rate at which data can be transmitted to Earth, just 1–4 kilobits per second. That's why it took over a year to get all the science data back to Earth. The team created special software to keep track of all the data sets and schedule when they were sent to Earth.

"The New Horizons mission has required patience for many years," said Weaver, "but we know the results will be well worth the wait."

The images of Pluto's 'heart' remain as some of the most intriguing of the mission.

"The large, bright, flat heart-shaped area is actually a giant sheet of molecular nitrogen ice," Stern explained. "We informally call it Sputnik Planum after the first robotic space explorer [the USSR's Sputnik satellite launched in 1957]. Along the periphery are mountain ranges, with elevations about as high as the Rockies in Colorado."

The mountains can't be made of nitrogen like the flat plain, Stern said, as nitrogen isn't strong enough to form mountains. They are probably water ice mountains, with a methane veneer on the surface.

One favorite of both Stern's and Weaver's is an image taken as New Horizons looked back at Pluto after its closest approach. "You could see Pluto's atmosphere!" Weaver said excitedly. "How many times can you say you've seen another planet's atmosphere? Plus it has all this structure in it."

The 'blue sky haze' in that picture has meaning for the team, too. "You can only take that picture from the farside of Pluto," Stern said, "meaning we had a successful flyby. For us, this was like the *Earthrise* picture taken by the Apollo astronauts."

Other images of Pluto's atmosphere strongly suggest cloud formations. If confirmed, they'd be the first clouds ever to be seen on the dwarf planet, and a sign this small world possesses an even more complex atmosphere than imagined.

A panoramic view of Pluto's surface from the Ralph instrument, a multispectral imager and infrared mapping spectrometer, is also striking. "That was mind-blowing," said Weaver. "I felt like I was there to see these mountains on Pluto's surface."

New Horizons also spotted what is likely a frozen former lake of liquid nitrogen located in a mountain range just north of Sputnik Planum. Liquids might have flowed across and pooled on the surface when Pluto had a higher atmospheric pressure with slightly warmer conditions.

New Horizons captured a near-sunset view of the rugged, icy mountains and flat ice plains extending to Pluto's horizon. The smooth expanse of the informally named Sputnik Planum (right) is flanked to the west (left) by rugged mountains up to 11,000 feet (3,500 m) high. More than a dozen layers of haze can be seen in Pluto's tenuous atmosphere. The image was taken from a distance of 11,000 miles (18,000 km) to Pluto; the scene is 230 miles (380 km) across. Credit: NASA/ JHUAPL/SwRI

Additionally, Pluto has floating water ice hills. Because water ice is less dense than nitrogen-dominated ice, scientists believe these water ice hills are floating in a sea of frozen nitrogen and move over time like icebergs in Earth's Arctic Ocean. The hills are likely fragments of the rugged uplands that have broken away and are being carried by the nitrogen glaciers into Sputnik Planum.

"We see that Pluto is geologically alive on a vast scale," Stern said. "Nothing like this has been seen anywhere in the solar system except where there are sources of tidal energy, such as the moons of Saturn and Jupiter. But this is a planet isolated out in space, and it is still massively active, with volcanoes on its surface. Sputnik Planum looks as though it was born yesterday, in geological terms."

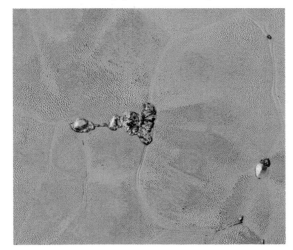

Sterns says in public talks he's been calling Pluto a sci-fi planet.

For decades, we could only imagine what the view of Pluto's surface might be. Now, we have the real thing.

High-resolution images of Pluto from New Horizons show Pluto's varied terrain—revealing details down to scales of 270 meters. In this 75-mile (120-km) section, the textured surface of the plain surrounds isolated ice mountains (which look like a Klingon Bird of Prey from Star Trek). Credit: NASA/JHUAPL/SWRI

Artist's impression of NASA's New Horizons spacecraft encountering a Pluto-like object in the distant Kuiper Belt. Credits: NASA/JHUAPL/SwRI/Alex Parker

"We've had artists' impressions for years," Weaver said, "based on what the Pluto Mafia had learned. Scientists like Marc Buie, John Spencer and Leslie Young had their scientifically based ideas, and they'd argue with Alan about what the colors would be. Since the 1980s, they knew Pluto was interesting. But they didn't know it was going to be as remarkable as what we've seen from New Horizons."

It wasn't only the spacecraft and the team that delivered, Weaver said, Pluto delivered, too. "It didn't have to be this spectacular, but it is."

THE MISSION CONTINUES

New Horizons continues on its blazing journey through the Kuiper Belt. It will continue scanning the region and the team wants to take images of at least twenty other distant KBOs, doing studies of their surface properties that can't be done from Earth.

How long can New Horizons keep going? NASA approved a mission extension to 2021, but the team hopes the mission can continue even longer.

"We believe we have enough power to operate the spacecraft and return scientific data until the mid-2030s," Bowman said. "The one limiting factor is the transmitter, which is the highest power consumer, taking about 32 watts to turn it on. So, assuming nothing else on the spacecraft fails, when we reach the point where we don't have power for the transmitter, that might end the mission."

However, NASA would have to agree to fund the mission that long.

But in the short term, New Horizons has its sights on another KBO named 2014 MU69. This icy object is a billion miles (1.6 billion km) beyond Pluto, and is about 30 miles (45 km) wide. New Horizons will fly past it for a close encounter on January 1, 2019, and the team hopes to bring the spacecraft even closer to this object than it was to Pluto.

Scientists suspect MU69 represents a primordial object that is a completely different class of KBO than Pluto.

"2014 MU69 is an ancient KBO, formed where it orbits now," Stern said. "It's the type of object scientists have been hoping to study for decades. This will be the most distant world we've ever been able to see up close, and it will be another mystery that New Horizons will unwrap for us."

CHAPTER 2

ROVING MARS WITH CURIOSITY

SEVEN MINUTES OF TERROR

It takes approximately seven minutes for a moderate-size spacecraft—such as a rover or a robotic lander—to descend through the atmosphere of Mars and reach the planet's surface. During those short minutes, the spacecraft has to decelerate from its blazing incoming speed of about 13,000 miles per hour (20,922 kph) to touch down at just 2 miles per hour (3 kph) or less.

This requires a Rube Goldberg–like series of events to take place in perfect sequence with precise choreography and timing. And it all needs to happen automatically via computer, with no input from anyone on Earth. There is no way to guide the spacecraft remotely from our planet, about 150 million miles (240 million km) away. At that distance, the radio signal delay time from Earth to Mars takes over thirteen minutes. Therefore, by the time the seven-minute descent is finished, all those events have happened—or not happened—and no one on Earth knows which. Either your spacecraft sits magnificently on the surface of Mars or lies in a crashed heap.

That's why scientists and engineers from the missions to Mars call it seven minutes of terror.

And with the Mars Science Laboratory (MSL) mission, which launched from Earth in November of 2011, the fear and trepidation about what is officially called the **entry, descent and landing** (EDL) increased exponentially. MSL features a one-ton (907 kg), six-wheeled rover named Curiosity, and this rover was going to use a brand-new, untried landing system.

An artist's concept depiction of the moment the Curiosity rover touches down onto the Martian surface, suspended on a bridle beneath the spacecraft's descent stage. Credit: NASA/JPL-Caltech

To date, all Mars landers and rovers have used—in order—a rocket-guided entry, a heat shield to protect and slow the vehicle, then a parachute followed by thrusters to slow the vehicle even more. Curiosity would use this sequence as well. However, a final, crucial component encompassed one of the most complex landing devices ever flown.

Dubbed the sky crane, this hovering rocket stage would lower the rover on 66-foot (20-m) cables of Vectran rope like a rappelling mountaineer, with the rover soft landing directly on its wheels. This all needed to be completed in a matter of seconds, and when the onboard computer sensed touchdown, pyrotechnics would sever the ropes and the hovering descent stage would zoom away at full throttle to crash-land far from Curiosity.

Complicating matters even further, this rover was going to attempt the most precise off-world landing ever, setting down inside a crater next to a mountain the height of Mount Rainier.

A major part of the uncertainty was that engineers could never test the entire landing system all together, in sequence. And nothing could simulate the brutal atmospheric conditions and lighter gravity present on Mars except being on Mars itself. Since the real landing would be the first time the full-up sky crane would be used, there were questions: What if the cables didn't separate? What if the descent stage kept descending right on top of the rover?

If the sky crane didn't work, it would be game over for a mission that had already overcome so much: technical problems, delays, cost overruns, and the wrath of critics who said this $2.5 billion Mars rover was bleeding money away from the rest of NASA's planetary exploration program.

Artist concept of one of the Viking landers on Mars. Credit: NASA/JPL

MISSIONS TO MARS

With its red glow in the nighttime sky, Mars has beckoned sky watchers for centuries. As the closest planet to Earth that offers any potential for future human missions or colonization, Mars has been of great interest in the age of space exploration. To date, over 40 robotic missions have been launched to the red planet . . . or, more precisely, 40-plus missions have been *attempted*.

Including all U.S., European, Soviet/Russian and Japanese efforts, more than half of the Mars missions have failed either because of a launch disaster, a malfunction en route to Mars, a botched attempt to slip into orbit or a catastrophic landing. While recent missions have had greater success than our first pioneering attempts to explore Mars in situ (on location), space scientists and engineers are only partially kidding when they talk about things like a "great galactic ghoul" or the 'Mars' curse messing up the missions.

But there have been wonderful successes, too. Early missions in the 1960s and 70s such as Mariner orbiters and Viking landers showed us a strikingly beautiful, although barren and rocky world, thereby dashing any hopes of little green men as our planetary neighbors. But later missions revealed a dichotomy: magnificent desolation combined with tantalizing hints of past— or perhaps even present day—water and global activity.

Today, Mars' surface appears cold and dry, and its whisper-thin atmosphere doesn't shield the planet from bombardment of radiation from the Sun. But indications are the conditions on Mars weren't always this way. Visible from orbit are channels and intricate valley systems that appear to have been carved by flowing water.

For decades, planetary scientists have debated whether these features formed during brief wet periods caused by cataclysmic events such as a massive asteroid strike or sudden climate calamity, or if they formed over millions of years when Mars may have been continuously warm and wet. Much of the evidence so far is ambiguous; these features could have formed either way. But billions of years ago, if there were rivers and oceans, just like on Earth, life might have taken hold.

THE ROVERS

The Curiosity rover is the fourth mobile spacecraft NASA has sent to Mars' surface. The first was a 23-pound (10.6-kg) rover named Sojourner that landed on a rock-covered Martian plain on July 4, 1997. About the size of a microwave oven, the 2-foot (65-cm)-long Sojourner never traversed more than 40 feet (12 m) away from its lander and base station. The rover and lander together constituted the Pathfinder mission, which was expected to last about a week. Instead, it lasted nearly three months, and the duo returned 2.6 gigabits of data, snapping more than 16,500 images from the lander and 550 images from the rover, as well as taking chemical measurements of rocks and soil and studying Mars' atmosphere and weather. It identified traces of a warmer, wetter past for Mars.

The mission took place when the Internet was just gaining popularity, and NASA decided to post pictures from the rover online as soon as they were beamed to Earth. This ended up being one of the biggest events in the young Internet's history, with NASA's website (and mirror sites set up for the high demand) receiving over 430 million hits in the first twenty days after landing.

Pathfinder, too, utilized an unusual landing system. Instead of using thrusters to touch down on the surface, engineers concocted a system of giant airbags to surround and protect the spacecraft. After using the conventional system of a rocket-guided entry, heat shield, parachutes and thrusters, the airbags inflated and the cocooned lander was dropped from 100 feet (30 m) above the ground. Bouncing several times across Mars' surface like a giant beach ball, Pathfinder eventually came to a stop, the airbags deflated and the lander opened up to allow the rover to emerge.

While that may sound like a crazy landing strategy, it worked so well that NASA decided to use larger versions of the airbags for the next rover mission: two identical rovers named Spirit and Opportunity. The Mars Exploration Rovers (MER) are about the size of a riding lawn mower, at 5.2 feet (1.6 m) long, weighing about 400 pounds (185 kg). Spirit landed successfully near Mars' equator on January 4, 2004, and three weeks later Opportunity bounced down on the other side of the planet. The goal of the MERs was to find evidence of past water on Mars, and both rovers hit the jackpot. Among many findings, Opportunity found ancient rock outcrops that were formed in flowing water, and Spirit found unusual cauliflower-shaped silica rocks that scientists are still studying, but they may provide clues to potential ancient Martian life.

Three Generations of Mars Rovers together at the "Mars Yard" at the Jet Propulsion Laboratory, which is a simulated Martian landscape used by the research and flight teams to test procedures for current rovers and prototypes for new missions: front, flight spare for the first Mars rover, Sojourner. Left, Mars Exploration Rover Project test rover. Right; Mars Science Laboratory test rover Credit: NASA/JPL-Caltech

A view of Eagle Crater on Mars where the Opportunity rover landed. Visible is the deflated landing airbags (white object in the center of the crater), tracks left by the rover and part of the solar arrays (bottom left). Credit: NASA/JPL-Caltech

Left Top: A panoramic view of Pathfinder's landing site at Ares Vallis on Mars. The Pathfinder rover, Sojourner, is shown snuggled against a rock. Credit: NASA/JPL

Left Bottom: Engineers test huge multi-lobed air bags created to protect the Mars Pathfinder Spacecraft as it reaches Mars' surface. The bags measure 17 feet (5 m) tall and 17 feet (5 m) in diameter. Credit: NASA/JPL

Incredibly, at this writing the Opportunity rover is still operating, driving more than a marathon (26 miles [42 km]), and it continues to explore Mars at a large crater named Endeavour. Spirit, however, succumbed to a loss of power during the cold Martian winter in 2010 after getting stuck in a sand trap. The two rovers far outlived their projected 90-day lifetime.

Somehow, the rovers each developed a distinct "personality"—or perhaps a better way to phrase it is that people *assigned* personalities to the robots. Spirit was a problem child and drama queen but had to struggle for every discovery. Opportunity was a privileged younger sister and star performer; new findings seemed to come easy for her. Spirit and Opportunity weren't designed to be adorable, but the charming rovers captured the imaginations of children and seasoned space veterans alike. MER project manager John Callas once called the twin rovers "the cutest darn things out in the solar system." As the long-lived, plucky rovers overcame hazards and perils, they sent postcards from Mars every day. And Earthlings loved them for it.

CURIOSITY

While it's long been on our space to-do list, we haven't quite yet figured out how to send humans to Mars. We need bigger and more advanced rockets and spacecraft, better technology for things like life support and growing our own food, and we really don't have the ability at this point to land the very large payloads needed to create a human settlement on Mars.

But in the meantime—while we try to figure all that out—we have sent the robotic equivalent of a human geologist to the red planet. The car-size Curiosity rover is armed with an array of seventeen cameras, a drill, a scoop, a hand lens and even a laser. These tools resemble equipment geologists use to study rocks and minerals on Earth. Additionally, this rover mimics human activity by mountain climbing, eating (figuratively speaking), flexing its (robotic) arm and taking selfies.

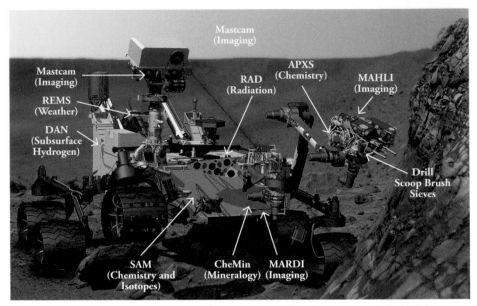

This graphic shows the locations of the cameras and instruments on NASA's Curiosity rover. Credit: NASA/JPL-Caltech

This roving robotic geologist is also a mobile chemistry lab. A total of ten instruments on the rover help search for organic carbon that might indicate the raw material required for life and "sniff" the Martian air, trying to smell if gases like methane—which could be a sign of life—are present. Curiosity's robotic arm carries a Swiss Army knife of gadgets: a magnifying lenslike camera, a spectrometer to measure chemical elements and a drill to bore inside rocks and feed samples to the laboratories named SAM (Sample Analysis at Mars) and CheMin (Chemistry and Mineralogy). The ChemCam laser can vaporize rock from up to 23 feet (7 m) away and identify the minerals from the spectrum of light emitted from the blasted rock. A weather station and radiation monitor round out the devices on board.

With these cameras and instruments, the rover becomes the eyes and hands for an international team of about 500 earthbound scientists.

While the previous Mars rovers used **solar arrays** to gather sunlight for power, Curiosity uses an RTG like New Horizons. The electricity generated from the RTG repeatedly powers rechargeable lithium-ion batteries, and the RTG's heat is also piped into the rover chassis to keep the interior electronics warm.

With Curiosity's size and weight, the airbag landing system used by the previous rovers was out of the question. As NASA engineer Rob Manning explained, "You can't bounce something that big." The sky crane is an audacious solution.

Curiosity's mission: figure out how Mars evolved over billions of years and determine if it once was—or even now is—capable of supporting microbial life.

Curiosity's target for exploration: a 3.4-mile (5.5-km)-high Mars mountain that scientists call Mount Sharp (formally known as Aeolis Mons) that sits in the middle of Gale Crater, a 96-mile (155-km)-diameter impact basin.

Top: Among those on hand for the landing include NASA Program Executive for Solar System Exploration Dave Lavery, second from right, and John Grunsfeld, the associate administrator of NASA's Science Mission Directorate, fourth from the right. Credit: NASA/Bill Ingalls. Bottom: MSL Flight Engineer Bobak Ferdowsi, right, caused an Internet sensation because of his stars-and-stripes Mohawk haircut, and became known as NASA's "Mohawk Guy." Credit: Ustream

Gale was chosen from 60 candidate sites. Data from orbiting spacecraft determined the mountain has dozens of layers of sedimentary rock, perhaps built over millions of years. These layers could tell the story of Mars' geological and climate history. Additionally, both the mountain and the crater appear to have channels and other features that look like they were carved by flowing water.

The plan: MSL would land in a lower, flatter part of the crater and carefully work its way upward toward the mountain, studying each layer, essentially taking a tour of the epochs of Mars' geological history.

The hardest part would be getting there. And the MSL team only had one chance to get it right.

LANDING NIGHT

Curiosity's landing on August 5, 2012, was one of the most anticipated space exploration events in recent history. Millions of people watched the events unfold online and on TV, with social media feeds buzzing with updates. NASA TV's feed from the Jet Propulsion Laboratory's (JPL's) mission control was broadcast live on the screens in New York's Time Square and at venues around the world hosting landing parties.

But the epicenter of action was at JPL, where hundreds of engineers, scientists and NASA officials gathered at JPL's Space Flight Operations Facility. The EDL team—all wearing matching light blue polo shirts—monitored computer consoles at mission control.

Two members of the team stood out: EDL team lead Adam Steltzner—who wears his hair in an Elvis-like pompadour—paced back and forth between the rows of consoles. Flight director Bobak Ferdowsi sported an elaborate stars and stripes Mohawk. Obviously, in the twenty-first century, exotic hairdos have replaced the 1960's black glasses and pocket protectors for NASA engineers.

At the time of the landing, Ashwin Vasavada was one of the longest-serving scientists on the mission team, having joined MSL as the deputy project scientist in 2004 when the rover was under construction. Back then, a big part of Vasavada's job was working with the instrument teams to finalize the objectives of their instruments, and supervising technical teams to help develop the instruments and integrate them with the rover.

MSL Project Scientist Ashwin Vasavada with a full-scale model of the Curiosity rover. Credit: NASA/JPL

Each of the ten selected instruments brought a team of scientists, so with engineers, additional staff and students, there were hundreds of people getting the rover ready for launch. Vasavada helped coordinate every decision and modification that might affect the eventual science done on Mars. During the landing, however, all he could do was watch.

"I was in the room next door to the control room that was being shown on TV," Vasavada said. "For the landing there was nothing I could do except realize the past eight years of my life and my entire future was all riding on that seven minutes of EDL."

Plus, the fact that no one would know the real fate of the rover until thirteen minutes after the fact due to the radio delay time led to a feeling of helplessness for everyone at JPL.

"Although I was sitting in a chair," Vasavada added, "I think I was mentally curled up in the fetal position."

As Curiosity sped closer to Mars, three other veteran spacecraft already orbiting the planet moved into position to be able to keep an eye on the newcomer MSL as it transmitted information on its status. At first, MSL communicated directly to the DSN antennas on Earth.

To make telemetry from the spacecraft as streamlined as possible during EDL, Curiosity sent out 128 simple but distinct tones indicating when steps in the landing process were activated. Allen Chen, an engineer in the control room, announced each as they came: one sound indicated the spacecraft entered Mars' atmosphere; another signaled the thrusters fired, guiding the spacecraft toward Gale Crater. Tentative clapping and smiles came from the team at mission control at the early tones, with emotions increasing as the spacecraft moved closer and closer to the surface.

Partway through the descent, MSL went below the Martian horizon, putting it out of direct communication with Earth. But the three orbiters—Mars Odyssey, Mars Reconnaissance Orbiter and Mars Express—were ready to capture, record and relay data to the DSN.

This unique image shows the heat shield of the Curiosity rover as it falls away after being jettisoned during the descent to the surface of Mars. The picture was taken by the Mars Descent Imager instrument (MARDI) located on the lower part of the rover chassis and the image shows the 15-foot (4.5-m) diameter heat shield when it was about 50 feet (16 m) from the spacecraft, about three seconds after heat shield separation and about two and a half minutes before touchdown. Credit: NASA/JPL-Caltech/ Malin Space Science Systems (MSSS)

The tweet sent out by the Curiosity rover Twitter account following the successful landing. Credit: Curiosity Rover Twitter Feed

Seamlessly, the tones kept coming to Earth as each step of the landing continued flawlessly. The parachute deployed. The heat shield dropped away. A tone signaled that the descent stage carrying the rover let go of the parachute, another indicated powered flight and descent toward the surface. Another tone meant the sky crane began lowering the rover to the surface.

A tone arrived, indicating that Curiosity's wheels touched the surface, but even that didn't mean success. The team had to make sure the sky crane flyaway maneuver worked.

Then came the tone they were waiting for: "Touchdown confirmed," cheered Chen. "We're safe on Mars!"

Pandemonium and joy erupted in JPL's mission control, at the landing party sites, and on social media. It seemed the world celebrated together at that moment. Cost overruns, delays, all the negative things ever said about the MSL mission seemed to vanish with the triumph of landing.

"Welcome to Mars!" the director of the Jet Propulsion Laboratory, Charles Elachi, said at a press conference following the dramatic touchdown. "Tonight we landed, tomorrow we start exploring Mars. Our Curiosity has no limits."

"The seven minutes actually went really fast," said Vasavada. "It was over before we knew it. Then everybody was jumping up and down, even though most of us were still processing that it went so successfully."

That the landing went so well—indeed perfectly—may have actually shocked some of the team members at JPL. While they had rehearsed Curiosity's landing several times, remarkably, they were never able to land the vehicle in their simulations.

An image captured by Curiosity shortly after it landed on the Red Planet on Aug. 5, 2012 PDT, showing the rover's main science target, Mount Sharp. The rover's shadow can be seen in the foreground, and the dark bands beyond are dunes. Credit: NASA/JPL-CalTech

"We tried to rehearse it very accurately," Vasavada said, "so that everything was in sync—both the telemetry that we had simulated that would be coming from the spacecraft, along with real-time animations that had been created. It was a pretty complex thing, but it never actually worked. So the real, actual landing was the first time everything worked right."

Curiosity was programmed to immediately take pictures of its surroundings. Within two minutes of the landing, the first images were beamed to Earth and popped up on the viewing screens at JPL.

"We had timed the orbiters to fly over during the landing but didn't know for sure if their relay link would last long enough to get the initial pictures down," Vasavada said. "Those first pictures were fairly ratty because the protective covers were still on the cameras and the thrusters had kicked up a lot of dust on the covers. We couldn't really see it very well, but we still jumped up and down nevertheless because these were pictures from Mars."

Amazingly, one of the first pictures showed exactly what the rover had been sent to study.

"We had landed with the cameras basically facing directly at Mount Sharp," Vasavada said, shaking his head. "In the Hazcam (hazard avoidance camera) image, right between the wheels, we had this gorgeous shot. There was the mountain. It was like a preview of the whole mission, right in front of us."

LIVING ON MARS TIME

The landing occurred at 10:30 p.m. Pacific time. The MSL team had little time to celebrate, transitioning immediately to mission operations and planning the rover's first day of activity. The team's first planning meeting started at 1:00 a.m. and ended at about 8:00 a.m. The team had been up all night, with some of the team working nearly 40 hours straight. This was a rough beginning for the scientists and engineers who needed to live on "Mars time."

A day on Mars—called a **sol**—is 40 minutes longer than it is on Earth, and for the first 90 sols of the mission, the entire team worked in shifts around the clock to constantly monitor the newly landed rover. To operate on the same daily schedule as the rover meant a perpetually shifting sleep-wake cycle, where the MSL team members would alter their schedules by 40 minutes every day to stay in sync with the day and night schedules on Mars. If team members came into work at 9:00 a.m., the next day they'd come in at 9:40 a.m., and the next day at 10:20 a.m., and so on.

Those who have lived through "Mars time" say their bodies continually feel jet-lagged. Some people slept at JPL so as not to disrupt their family's schedule, some wore two watches so they would know what time it was on both planets. About 350 scientists from around the world were involved with MSL, and many of them stayed at JPL for the first 90 sols of the mission, living on Mars time.

But it took fewer than 60 Earth days for the team to announce Curiosity's first big discovery.

WATER, WATER . . .

Ashwin Vasavada grew up in California and has fond childhood memories of visiting state and national parks in the southwest United States with his family, playing among sand dunes and hiking in the mountains. He's now able to do both on another planet, vicariously through Curiosity. The day I visited Vasavada at his office at JPL in early 2016, the rover was navigating through a field of giant sand dunes at the base of Mount Sharp, with some dunes towering 30 feet (9 m) above the rover.

A 16-foot (5 m) high sand dune on Mars called Namib Dune is part of the dark-sand "Bagnold Dunes" field along the northwestern flank of Mount Sharp. Images taken from orbit indicate that dunes in the Bagnold field move as much as about 3 feet (1 m) per Earth year. This image is part of a 360 degree panorama taken by the Curiosity rover on Dec. 18, 2015 or the 1,197th Martian day, or sol, of the rover's work on Mars. Credit: NASA/JPL-Caltech/ MSSS

"It's just fascinating to see dunes close up on another planet," Vasavada said. "And the closer we get to the mountain, the more fantastic the geology gets. So much has gone on there, and we have so little understanding of it . . . as of yet."

At the time we talked, Curiosity was approaching four Earth years on Mars. The rover is now studying those enticing sedimentary layers on Mount Sharp in closer detail. But first, it needed to navigate through the Bagnold Dunes, which form a barrier along the northwestern flank of the mountain. Here, Curiosity was doing what Vasavada calls flyby science, stopping briefly to sample and study the sand grains of the dunes while moving through the area as quickly as possible.

Now working as the lead project scientist for the mission, Vasavada plays an even larger role in coordinating the mission.

"It's a constant balance of doing things quickly, carefully and efficiently, as well as using the instruments to their fullest," he said.

Since the successful August 2012 landing, Curiosity has sent back tens of thousands of images from Mars—from expansive panoramas to extreme close-ups of rocks and sand grains, all of which are helping to tell the story of Mars' past.

The images the public seems to love the most are the selfies, the photos the rover takes of itself sitting on Mars. The selfies aren't just a single image like the ones we take with our cell phones but a mosaic created from dozens of separate images taken with the Mars Hand Lens Imager (MAHLI) camera at the end of the rover's robotic arm. Other fan favorites are the pictures Curiosity takes of the magnificent Martian landscape, like a tourist documenting its journey.

The rippled surface of the first Martian sand dune ever studied up close fills this Nov. 27, 2015, view of "High Dune" from the Mast Camera on NASA's Curiosity rover. This site is part of the "Bagnold Dunes" field of active dark dunes along the northwestern flank of Mount Sharp. Credit: NASA/JPL-Caltech/MSSS

Vasavada has a unique personal favorite.

"For me, the most meaningful picture from Curiosity really isn't that great of an image," he said, "but it was one of our first discoveries so it has an emotional tie to it."

Within the first 50 sols, Curiosity took pictures of what geologists call conglomerates: a rock made of pebbles cemented together. But these were no ordinary pebbles—they were pebbles worn by flowing water. Serendipitously, the rover had found an ancient streambed where water once flowed vigorously. From the size of the pebbles, the science team could interpret the water was moving about three feet (1 m) per second, with a depth somewhere between a few inches to several feet.

"When you see this picture, and whether you are a gardener or geologist, you know what this means," Vasavada said excitedly. "At Home Depot, the rounded rocks for landscaping are called river pebbles! It was mind-blowing to me to think that the rover was driving through a streambed. That picture really brought home there was actually water flowing here long ago, probably ankle to hip deep." Vasavada looked down. "It still gives me the shivers, just thinking about it," he said, with his passion for exploration and discovery visibly evident.

"Selfies" taken by the Curiosity rover are actually a mosaic created from numerous images taken with the Mars Hand Lens Imager (MAHLI), located on the end of the rover's robotic arm. However, the arm is not shown in the selfies, because with the wrist motions and turret rotations used in pointing the camera for the component images, the arm was positioned out of the shot in the frames or portions of frames used, just like your hand and arm are behind the camera when you take a picture. However, the shadow of the arm is visible on the ground. This low-angle selfie shows the vehicle at the site from which it reached down to drill into a rock target called "Buckskin" on lower Mount Sharp. Credit: NASA/JPL-Caltech/MSSS

This geological feature on Mars is exposed bedrock made up of smaller fragments cemented together, or what geologists call a sedimentary conglomerate, and is evidence for an ancient, flowing stream. Some of embedded and loose gravel are round in shape, leading the Curiosity science team to conclude it was transported by a vigorous flow of water. Curiosity's 100-millimeter Mastcam telephoto lens took this picture on the 39th sol of the mission (Sept. 14, 2012). Credit: NASA/JPL-Caltech/MSSS

5 cm

From that early discovery, Curiosity continued to find more water-related evidence. The team took a calculated gamble and instead of driving straight toward Mount Sharp took a slight detour to the east to an area dubbed Yellowknife Bay.

"Yellowknife Bay was something we saw with the orbiters," Vasavada explained, "and there appeared to be a debris fan fed by a river—evidence of flowing water in the ancient past."

Here, Curiosity fulfilled one of its main goals: determining whether Gale Crater ever was habitable for simple life forms. The answer was a resounding yes. The rover sampled two stone slabs with the drill, feeding portions the size of half a baby aspirin to SAM, the onboard lab. SAM identified traces of elements like carbon, hydrogen, nitrogen, oxygen and more—the basic building blocks of life. It also found sulfur compounds in different chemical forms, a possible energy source for microbes.

Data gathered by Curiosity's other instruments constructed a portrait detailing how this site was once a muddy lake bed with mild—not acidic—water. Add in the essential elemental ingredients for life, and long ago Yellowknife Bay would have been the perfect spot for living organisms to hang out. While this finding doesn't necessarily mean there was past (or present) life on Mars, it shows that the raw ingredients existed for life to get started there at one time, in a benign environment.

"Finding the habitable environment in Yellowknife Bay was wonderful because it really showed the capability our mission has to measure so many different things," Vasavada said. "A wonderful picture came together of streams that flowed into a lake environment. This was exactly what we were sent there to find, but we didn't think we'd find it that early in the mission."

Still, this lake bed could have been created by a one-time event over just hundreds of years. The jackpot would be to find evidence of long-term water and warmth.

That discovery took a little longer. But personally, it means more to Vasavada.

Mars' climate was one of Vasavada's early interests in his career, and he spent years creating models, trying to understand Mars' ancient history.

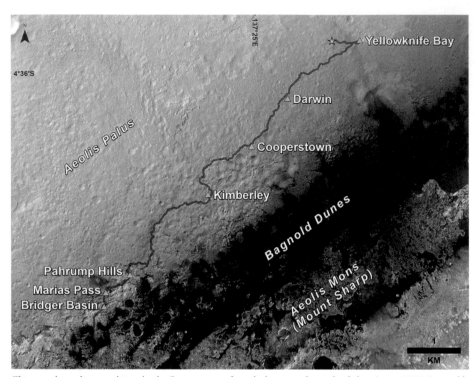

This map shows the route driven by the Curiosity rover from the location where it landed in August 2012 (notated by the star) to its location in December 2015, showing investigation targets such as Yellowknife Bay, Pahrump Hills and the Bagnold Dunes. Credit: NASA/JPL-Caltech/Univ. of Arizona

"I grew up with pictures of Mars from the Viking mission," he said, "and thinking of it as a barren place with jagged volcanic rock and a bunch of sand. Then I had done all this theoretical work about Mars' climate, that rivers and oceans perhaps once existed on Mars, but we had no real evidence."

That's why the discovery made by Curiosity in late 2015 is so exciting to Vasavada and his team.

"We didn't just see the rounded pebbles and remnants of the muddy lake bottom at Yellowknife Bay, but all along the route," Vasavada said. "We saw river pebbles first, then tilted sandstones where the river emptied into a lake. Then as we got to Mount Sharp, we saw huge expanses of rock made of the silt that settled out from the lake."

The explanation that best fits the morphology in this region—that is, the configuration and evolution of rocks and landforms—is that rivers formed deltas as they emptied into a lake. This likely occurred 3.8 to 3.3 billion years ago. And the rivers delivered sediment that slowly built up the lower layers of Mount Sharp.

"My gosh, we were seeing this full system now," Vasavada explained, "showing how the entire lower few hundred meters of Mount Sharp were likely laid down by these river and lake sediments. That means this event didn't take hundreds or thousands of years; it required millions of years for lakes and rivers to be present to slowly build up, millimeter by millimeter, the bottom of the mountain."

Layers at the base of Mt. Sharp. These visible layers in Gale Crater show the chapters of the geological history of Mars in this image from NASA's Curiosity rover. The image shows the base of Mount Sharp, the rover's science destination, and was taken with Curiosity's Mast Camera on Aug. 23, 2012. Credit: NASA/JPL-Caltech/MSSS

For that, Mars also needed a thicker atmosphere than it has now, and a greenhouse gas composition that Vasavada said they haven't quite figured out yet.

But then, somehow dramatic climate change caused the water to disappear and winds in the crater carved the mountain to its current shape.

The rover had landed in exactly the right place because here in one area was a record of much of Mars' environmental history, including evidence of a major shift in the planet's climate, when the water that once covered Gale Crater with sediment dried up.

"This all is a significant driver now for what we need to explain about Mars' early climate," Vasavada said. "You don't get millions of years of climate change from a single event like a meteor hit. This discovery has broad implications for the entire planet, not just Gale Crater."

OTHER DISCOVERIES

- **Silica:** The rover made a completely unanticipated discovery of high-content silica rocks as it approached Mount Sharp. "This means that the rest of the normal elements that form rocks were stripped away, or that a lot of extra silica was added somehow," Vasavada said, "both of which are very interesting and very different from rocks we had seen before. It's such a multifaceted and curious discovery, we're going to take a while figuring it out."

- **Methane:** Methane is usually a sign of activity involving organic matter—even, potentially, of life. On Earth, about 90 percent of atmospheric methane is produced from the breakdown of organic matter. On Mars, methane has been detected by other missions and telescopes over the years, but it was tenuous—the readings seemed to come and go and are hard to verify. In 2014, the Tunable Laser Spectrometer within the SAM instrument observed a ten-fold increase in methane over a two-month period. What caused the brief and sudden increase? Curiosity will continue to monitor readings of methane, and hopefully provide an answer to the decades-long debate.

- **Radiation Risks for Human Explorers:** Both during its trip to Mars and on the surface, Curiosity measured the high-energy radiation from the Sun and space that poses a risk to astronauts. NASA will use data from the Radiation Assessment Detector (RAD) instrument to help design future missions to be safe for human explorers.

HOW TO DRIVE A MARS ROVER

How does Curiosity know where and how to drive across Mars' surface? You might envision engineers at JPL using joysticks, similar to those used for remote control toys or video games. But unlike RC driving or gaming, the Mars rover drivers don't have immediate visual inputs or a video screen to see where the rover is going. And just as with the landing, there is always a time delay between when a command is sent to the rover and when it is received on Mars.

"It's not driving in a real-time interactive sense because of the time lag," explained John Michael Morookian, who leads the team of rover drivers.

The actual job title of Morookian and his team is rover planner, which precisely describes what they do. Instead of driving the rovers per se, they plan out the route in advance, then program specialized software, and upload the instructions to Curiosity.

"We use images taken by the rover of its surroundings," said Morookian. "We have a set of stereo images from four black-and-white navigation cameras, along with images from the Hazcams supported by high-resolution color images from the Mastcam that give us details about the nature of the terrain ahead and clues about types of rocks and minerals that are at the site. This helps identify structures that look interesting to the scientists."

Using all available data, they can create a three-dimensional visualization of the terrain with specialized software called the **Rover Sequencing and Visualization Program** (RSVP).

"This is basically a Mars simulator, and we put a simulated Curiosity in a panorama of the scene to visualize how the rover could traverse on its path," Morookian explained. "We can also put on stereo glasses, which allow our eyes to see the scene in three dimensions as if we were there with the rover."

In virtual reality, the rover drivers can manipulate the scene and the rover to test every possibility of which routes are the best and what areas to avoid. There, they can make all the mistakes (get stuck in a dune, tip the rover, crash into a big rock, drive off a precipice) and perfect the driving sequence while the real rover remains safe on Mars.

"The scientists also review the images for features that are interesting and consult with the rover planners to help define a path. Then we compose the detailed commands that are necessary to get Curiosity from point A to point B along that path," Morookian said. "We can also incorporate the commands needed to give the rover direction to make contact with the site using its robotic arm."

So every night the rover is commanded to shut down for eight hours to recharge its batteries with the nuclear generator. But first, Curiosity sends data to Earth, including pictures of the terrain and any science information. On Earth, the rover planners take that data, do their planning work, complete the software programming and beam the information back to Mars. Then Curiosity wakes up, downloads the instructions and sets to work. And the cycle repeats.

Curiosity also has an autonav feature that allows the rover to traverse areas the team hasn't seen yet in images. So it could go over the hill and down the other side to uncharted territory, with the autonav sensing potential hazards.

This self-portrait of NASA's Curiosity Mars rover shows the vehicle at the 'Big Sky' site, where it drilled into sandstone rock, at lower left corner. The scene combines images taken by the Mars Hand Lens Imager (MAHLI) camera on Sol 1126 (Oct. 6, 2015). Credit: NASA/JPL-Caltech/MSSS

"We don't use it too often because it is computationally expensive, meaning it takes much longer for the rover to operate in that mode," Morookian said. "We often find it's a better trade to just come in the next day, look at the images and drive as far as we can see."

As Morookian showed me the various rooms used by rover planning teams at JPL, he explained how they need to operate over a number of different timescales.

"We not only have the daily route planning," he said, "but also do long-range strategic planning using orbital imagery from the HiRISE camera on the Mars Reconnaissance Orbiter (read about that mission on page 161) and choose paths based on features seen from orbit. Our team works strategically, looking many months out to define the best paths."

Another process called **supratactical** looks out to just the next week. This involves science planners managing and refining the types of activities the rover will be doing in the short term. Also, since no one on the team lives on Mars time anymore, on Fridays the rover planners work out the plans for several days.

"Since we don't work weekends, Friday plans contain multiple sols of activities," Morookian said. "Two parallel teams decide which days the rover will drive and which days it will do other activities, such as work with the robotic arm or other instruments."

The data that comes down from the rover over the weekend is monitored, however, and if there is a problem, a team is called in to do a more detailed assessment. Morookian indicated they've had to engage the emergency weekend team several times, but so far there have been no serious problems. "It does keep us on our toes, however," he said.

The rover features a number of reactive safety checks, such as monitoring the amount of overall tilt of the rover deck and the articulation of the suspension system of the wheels, so if the rover is going over an object that is too large, it will automatically stop.

When Curiosity's Navigation Cameras (Navcams) take black-and-white images and send them back to Earth each day, rover planners combine them with other rover data to create 3D terrain models. By adding a computerized 3D rover model to the terrain model, rover planners can understand better the rover's position, as well as distances to, and scale of, features in the landscape. Credit: NASA/JPL-Caltech

This image shows a close-up of track marks left by the Curiosity rover. Holes in the rover's wheels, seen here in this view, leave imprints in the tracks that can be used to help the rover drive more accurately. The imprint is Morse code for 'JPL,' and aids in tracking how far the rover has traveled. Credit: NASA/JPL-Caltech

Curiosity wasn't built for speed. It was designed to travel up to 660 feet (200 m) in a day, but it rarely travels that far in a sol. By mid 2016, the rover had driven a total of about 8.2 miles (13.2 km) across Mars' surface.

There are several ways to determine how far Curiosity has traveled, but the most accurate measurement is called **visual odometry**. Curiosity has specialized holes in its wheels in the shape of Morse code letters, spelling out *JPL*—a nod to the home of the rover's science and engineering teams—across the Martian soil.

"Visual odometry works by comparing the most recent pair of stereo images collected roughly every meter over the drive," said Morookian. "Individual features in the scene are matched and tracked to provide a measure of how the camera (and thus the rover) has translated and rotated in three-dimensional space between the two images, and it tells us in a very real sense how far Curiosity has gone."

Careful inspection of the rover tracks can reveal the type of traction the wheels have and if they have slipped, for instance due to high slopes or sandy ground.

Unfortunately, Curiosity now has new holes in its wheels that aren't supposed to be there.

ROVER PROBLEMS

Morookian and Vasavada both expressed relief and satisfaction that overall—this far into the mission—Curiosity is a fairly healthy rover. The entire science payload is currently operating at nearly full capability. But the engineering team keeps an eye on a few issues.

"Around sol 400, we realized the wheels were wearing faster than we expected," Vasavada said.

And the wear didn't consist of just little holes; the team started to see punctures and nasty tears. Engineers realized the holes were being created by the hard, jagged rocks the rover was driving over during that time.

"We weren't fully expecting the kind of 'pointy' rocks that were doing damage," Vasavada said. "We also did some testing and saw how one wheel could push another wheel into a rock, making the damage worse. We now drive more carefully and don't drive as long as we have in the past. We've been able to level off the damage to a more acceptable rate."

The team operating the Curiosity Mars rover uses the Mars Hand Lens Imager (MAHLI) camera on the rover's arm to check the condition of the wheels at routine intervals. This image of Curiosity's left-middle and left-rear wheels is part of an inspection set taken on April 18, 2016, during the 1,315th sol of the rover's work on Mars, and shows unexpected holes in the wheels. Credit: NASA/JPL-Caltech/MSSS

Early in the mission, Curiosity's computer went into safe mode several times, which means when Curiosity's software recognizes a problem, the response is to disallow further activity and phone home.

Specialized fault protection software runs throughout the modules and instruments, and when a problem occurs, the rover stops and sends data called event records to Earth. The records include various categories of urgency, and in early 2015, the rover sent a message that essentially said, "This is very, very bad." The drill on the rover's arm had experienced a fluctuation in an electrical current—like a short circuit.

"Curiosity's software has the ability to detect shorts, like the ground fault circuit interrupter you have in your bathroom," Morookian explained, "except this one tells you 'this is very, very bad' instead of just giving you a yellow light."

Since the team can't go to Mars and repair a problem, everything is fixed either by sending software updates to the rover or by changing operational procedures.

"We are just more careful now with how we use the drill," Vasavada said, "and don't drill with full force at the beginning, but slowly ramp up. It's sort of like how we drive now, more gingerly but it still gets the job done. It hasn't been a huge impact as of yet."

A lighter touch on the drill also was necessary for the softer mudstones and sandstones the rover encountered. Morookian said there was concern the layered rocks might not hold up under the assault of the standard drilling protocol, so they adjusted the technique to use the lowest settings that still allow the drill to make sufficient progress into the rock.

But opportunities to use the drill are increasing as Curiosity begins its traverse up the mountain. The rover is traveling through what Vasavada calls a "target-rich, very interesting area," as the science team works to tie together the geological context of everything they are seeing in the images.

Curiosity's drill in the turret of tools at the end of the robotic arm positioned in contact with the rock surface for the first drilling of the mission on the 170th sol of Curiosity's work on Mars (Jan. 27, 2013) in Yellowknife Bay. The picture was taken by the front Hazard-Avoidance Camera (Hazcam). Credit: NASA/ JPL-Caltech

FINDING BALANCE ON MARS

While the diversion at Yellowknife Bay allowed the team to make some major discoveries, they felt pressure to get to Mount Sharp, so they "drove like hell for a year," Vasavada said.

Now on the mountain, there is still the pressure to make the most of the mission, with the goal of making it through at least four different rock units—or layers—on Mount Sharp. Each layer could be like a chapter in the book of Mars' history.

"Exploring Mount Sharp is fascinating," Vasavada said, "and we're trying to maintain a mix between really great discoveries, which—you hate to say—slows us down, and getting higher on the mountain. Looking closely at a rock in front of you means you'll never be able to go over and look at that other interesting rock over there."

Vasavada and Morookian both said it's a challenge to preserve that balance every day—to find what's called the knee in the curve or sweet spot of the perfect optimization between driving and stopping for science.

Then there's the balance between stopping to do a full observation with all the instruments or doing flyby science where less intense observations are made.

"We take the observations we can, and generate all the hypotheses we can in real time," Vasavada said. "Even if we're left with 100 open questions, we know we can answer the questions later as long as we know we've taken enough data."

Curiosity's primary target is not the summit of Mount Sharp but instead a region about 1,330 feet (400 m) up where geologists expect to find the boundary between rocks that saw a lot of water in their history and those that didn't. That boundary will provide insight into Mars' transition from a wet planet to dry, filling in a key gap in the understanding of the planet's history.

No one really knows how long Curiosity will last, or if it will surprise everyone like its predecessors Spirit and Opportunity. Having made it past the prime mission of one year on Mars (two Earth years), and now in the extended mission, the one big variable is the RTG power source. While the available power will start to steadily decrease, both Vasavada and Morookian don't expect that to be in an issue for at least four more Earth years, and with the right "nurturing," power could last for a dozen years or more.

But they also know there's no way to predict how long Curiosity will go, or what unexpected event might end the mission.

Curiosity appears to be photobombing Mount Sharp in this selfie image, a mosaic created from several MAHLI images. Credit: NASA/JPL-Caltech/MSSS/Edited by Jason Major

THE BEAST

Does Curiosity have a personality like the previous Mars rovers?

"Actually no, we don't seem to anthropomorphize this rover like people did with Spirit and Opportunity," Vasavada said. "We haven't bonded emotionally with it. Sociologists have actually been studying this." He shook his head with an amused smile.

Vasavada indicated it might have something to do with Curiosity's size.

"I think of it as a giant beast," he said straight-faced. "But not in a mean way at all."

What has come to characterize this mission, Vasavada said, is the complexity of it in every dimension: the human component of getting 500 people to work and cooperate together while optimizing everyone's talents, keeping the rover safe and healthy, and keeping ten instruments going every day, which are sometimes doing completely unrelated science tasks.

"Every day is our own little 'seven minutes of terror,' where so many things have to go right every single day," Vasavada said. "There are a million potential issues and interactions, and you have to constantly be thinking about all the ways things can go wrong, because there are a million ways you can mess up. It's an intricate dance, but fortunately we have a great team."

Then he added with a smile, "This mission is exciting, though, even if it's a beast."

CHAPTER 3

CHANGING EVERYTHING: THE REMARKABLE HUBBLE SPACE TELESCOPE

STARGAZING

Like a child standing on tiptoe to see farther, astronomers have always stretched and reached—moving their telescopes higher and higher to rooftops, hilltops and then mountaintops—to gain a better view of the heavens. Try as they might, no matter where they go on Earth, astronomers can't completely eliminate the effects of Earth's luxuriously thick atmosphere. While the majority of us appreciate its life-giving qualities, for anyone who looks through telescopes, the blanket of air that covers our planet is a curse.

Unstable air pockets can bend starlight in random directions, creating what's called **atmospheric distortion**. This causes stars to twinkle, which, again, most of us might like, but this distortion makes it difficult to clearly see stars and other astronomical objects. Techniques to mitigate the atmosphere's effects have been developed, but still, clouds and rain obviously deter telescope use. The atmosphere also partially blocks or absorbs certain wavelengths of light that astronomers find very interesting, like ultraviolet, gamma- and X-rays. However, these are also somewhat harmful to humans, so thanks again, atmosphere.

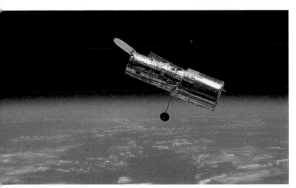

The Hubble Space Telescope hovers at the boundary of Earth and space in this picture, taken after Hubble's second servicing mission in 1997. Hubble orbits 343 miles (552 km) above the Earth's surface, where it can avoid the atmosphere and clearly see objects in space. Credit: NASA

One of the most iconic images from the Hubble Space Telescopes is the 'Pillars of Creation," which shows gas pillars in the Eagle Nebula. These eerie, pillar-like structures are actually columns of cool interstellar hydrogen gas and dust where new stars are born. The Eagle Nebula is about 6,500 light-years away from Earth, and the tallest pillar (left) is about 4 light-years long from base to tip. The picture was taken on April 1, 1995 with the Hubble Space Telescope Wide Field and Planetary Camera 2. Credit: NASA, ESA, STScI, J. Hester and P. Scowen (Arizona State University)

But since the dawn of the space age, astronomers realized they could solve all of these atmosphere-induced problems by putting a telescope on the ultimate mountaintop, where there's no atmosphere at all: space.

"I think everyone knew the Hubble Space Telescope would really revolutionize astronomy," said Helmut Jenkner, who is currently the interim head of the Hubble Space Telescope mission at the Space Telescope Science Institute (STScI) in Baltimore, Maryland. "But I don't think anyone expected it would still be such a productive, world-class observatory even more than a quarter century after it launched."

HST, as the telescope is affectionately called, wasn't the first telescope in space when it rocketed to Earth orbit in 1990, but many regard it as the best telescope ever, and undoubtedly the most well-known. Ask people to name a telescope and most will probably say, "Hubble." Walk into a classroom and you'll likely see a picture from HST on the wall. Name an astronomical conundrum and Hubble has probably studied it, maybe even solved it. This iconic space telescope has become one of the most successful space missions ever, both in terms of scientific return and its impact on the public.

But there have also been moments in HST's history when it appeared that the decades of effort and billions of dollars in taxpayer money might all have been for naught. At one point Hubble was considered a failure.

But Hubble's story is one of redemption and overcoming obstacles to provide spectacular views of the cosmos and reveal extraordinary insights about the universe.

"This mission has certainly been a roller coaster, it's had its ups and downs," said Ken Sembach, director of STScI, the home of 650 astronomers, and technical and administrative personnel who work on Hubble and other missions. "But HST always seems to have a way to overcome any hurdles, and I like to say that when things seemed really bad, that was OK because really, it was just an opportunity."

To be honest, Sembach says, Hubble is probably a far better observatory for being canceled, delayed and almost left for dead than if it all would have gone as originally planned.

OBSTACLES

Hubble's first obstacle came when trying to find enough money to build it. The U.S. Congress balked at a potential price tag of $400 million in the early 1970s and canceled what was then called the Large Space Telescope. But that provided an opportunity for the ESA to say it would like to get involved, bringing both cash and expertise.

"You may have noticed I have an accent," Jenkner said with his engaging smile, in his Arnold Schwarzenegger–like baritone. Originally from Austria, Jenkner joined the project in 1983 as an ESA representative. "Hubble is an international project and the inherent diversity in that approach has helped Hubble immensely, bringing in experts from around the world from every field. Walk through the halls of the Institute and you'll hear all sorts of accents."

Jenkner's first task was helping to compile a catalog of guide stars. HST utilizes what's called **Fine Guidance Sensors** that track stars with incredible accuracy to help the telescope find its way as it scans the heavens and keep Hubble locked onto to its targets. To work correctly, Hubble must be able to stay focused on a target without deviating more than 7/1,000ths of an arc second, or about the width of a human hair seen at a distance of a mile. To keep the telescope steady, HST has six gyroscopes and four free-spinning steering devices called **reaction wheels**.

Jenkner was the chief systems analyst for the guide star catalog project and along with a group of other young, tech-savvy-for-the-time astronomers and engineers, developed the software system that would allow Hubble to autonomously select a pair of stars in every possible direction so the telescope knew exactly where it was looking. This involved scanning nearly 1,500 glass plates to convert them to digital form.

Photography and astronomy were revolutionized in 1969 with the invention of the charge-coupled device (CCD). These image sensors are now in everything from the camera on your cell phone to the largest telescopes in space, including Hubble. But in the early 1980s, almost all of the historical astronomical data was still stored on glass photographic plates that had been used since the 1850s.

Helmut Jenkner (dark suit) greets Drew Feustel, John Grunsfeld and other astronauts from the 2009 Hubble servicing mission as they arrive at the Space Telescope Science Institute for a visit. Credit: NASA, ESA, and J. Coyle Studios for STScI

The assignment for Jenkner and colleagues was to create a detailed digital database from the glass plates, but it needed to be about 100 times larger than any previous star catalog and contain nearly twenty million objects.

"We were successful, but in retrospect I really regret that nobody told us it couldn't be done," Jenkner said with a laugh. "We were too young and too idealistic to know it was almost impossible. We were pushing the limits of what was possible in regards to data processing at the time."

Several other challenges presented themselves in building a complex telescope that required systems and procedures that had never been done before, including building huge sensitive mirrors (Hubble's primary mirror is eight feet [2.4 m] in diameter) that had to be perfectly shaped but also withstand the rigors of riding a rocket to space. Plus, HST would be launched from a space shuttle, so it had to meet certain safety requirements for human spaceflight. With the technical challenges came delays, and with delays came increased costs. But the team worked diligently toward a potential launch date in late 1986.

But then in January of '86, the space shuttle Challenger exploded shortly after liftoff, killing all seven astronauts aboard. The entire space shuttle program was put on hold, and since Hubble was supposed to be launched on Challenger's next mission, everyone knew it would be at least a couple of years until Hubble went into space. While this was a challenging time for the U.S. space program, it did provide an opportunity to refine and enhance Hubble's capabilities, especially the software for running the telescope from the ground.

Finally, Hubble rode to its destination in space on board Space Shuttle Discovery's STS-31 mission on April 24, 1990. Astronauts successfully deployed the telescope with a $1.5 billion price tag and all seemed well.

But shortly after HST was up and running, the science team realized the images being returned to Earth were blurry. It turned out the telescope's mirror had a **spherical aberration**, meaning it had been polished in slightly the wrong shape, off by just 2 microns. A sheet of paper is about 100 microns thick, so the flaw in Hubble's mirror was extremely small. But it was enough to give Hubble bad eyesight, distorting the images.

NASA, ESA and astronomers around the world were shocked and disappointed, the public was let down and politicians were furious. Hubble became the butt of jokes, a big white elephant in space.

The Space Shuttle Discovery launches with the Hubble Space Telescope on board. Credit: NASA

Astronauts work on the Hubble Telescope, attached to the Space Shuttle Endeavour's payload bay in 1993 during the first servicing mission. F. Story Musgrave is anchored on the Shuttle's robotic arm and Jeffrey Hoffman works below. Credit: NASA

MR. FIXIT IN SPACE

But, of course, Hubble's story doesn't end there. In fact, it was just the beginning.

Any other space telescope with a faulty mirror would have been declared dead. But the hope of fixing Hubble lay in the fact it was one of the very few space missions ever designed with modular instruments that could be swapped out and upgraded over time, placed in low Earth orbit, where space shuttle astronauts could perform the work. NASA already had a schedule of servicing missions planned for Hubble, and the first servicing mission could potentially provide an opportunity to fix the flawed telescope. But how? While the instruments could be changed or tweaked, the mirror could not. A worldwide effort ensued to figure out what could be done.

"It was a hair-raising, nerve-wracking time for everyone at NASA," said Frank Cepollina, who was the head of Hubble servicing at NASA's Goddard Space Flight Center in Greenbelt, Maryland. Cepollina had helped develop the concept of modular, fixable spacecraft, and he was in charge of planning and choreographing the servicing missions the astronauts would perform on Hubble. But when that first mission meant life or death for the telescope, Cepollina and his team switched into high gear. And they relished the challenge.

"While everyone else was antsy and nervous, for us, it was fun," Cepollina said with gusto, recalling the heady, adrenaline-filled days when NASA turned to him and his team to accomplish what seemed impossible. "It was a challenge, doing something that had never been done before. Really, it was ingenuity that won the day: the ingenuity to decide how to fix this 8-foot (2.4-m) diameter mirror with a tiny error in polishing, and the ingenuity too for us to figure out how to execute it."

The idea was to create a set of "eyeglasses" to fix Hubble's vision. Scientists, engineers and optical experts worked together to create a set of corrective lenses called **Corrective Optics Space Telescope Axial Replacement** (COSTAR) for three of Hubble's instruments.

"It was one thing to build COSTAR," said Cepollina, "but to position those 'glasses' correctly and to get the spacing just right for each instrument was really a technical nightmare. They had to design and build things and polish the corrective lenses to the finest quality, and, again, all this had never been done before."

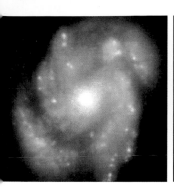
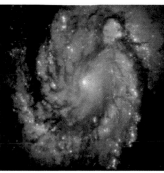

These comparison images of the galaxy M100 shows the dramatic improvement in the Hubble Space Telescope's view of the universe. On left is a picture taken with the original Wide Field and Planetary Camera 1 (WFPC-1) on November 27, 1993, just a few days prior to the first Hubble servicing mission, showing the effects of the flaw in HST's primary mirror which blurred starlight and limited the telescope's ability to see faint structure. The new image (right) was taken with the second generation WFPC-2 on December 31, 1993, demonstrating the corrective optics incorporated within the new camera compensating for the optical aberration, allowing Hubble to probe the universe with unprecedented clarity and sensitivity. Credit: NASA

The other challenge for Cepollina—now in his 80s and still active in developing robotics at Goddard—was to create the specialized tools astronauts would need to work in the zero-gravity environment of space, enabling them to do fine, detailed work while wearing bulky spacesuits, with the equivalent of big winter gloves on their hands.

After COSTAR was developed, Cepollina and his team spent more than a year training astronauts for one of the most complex space missions ever attempted. While the mission, dubbed Servicing Mission 1 (SM1), was critical for Hubble's future, if successful it also would be a huge boon to the human-robotic interface, where astronauts could go fix ailing satellites, which Cepollina had been touting for years.

During the ten-day servicing mission near the end of 1993, it took five grueling spacewalks for astronauts on space shuttle Endeavour to install COSTAR along with other equipment, including the updated Wide Field and Planetary Camera 2 (with its own corrective optics), new gyroscopes and solar panels. Each update was designed to fix or improve all the systems so Hubble would operate as originally intended, or perhaps even better. The world watched and waited to see if the astronauts' perilous work had done the trick.

In early January of 1994, NASA released the first new images from Hubble's fixed optics. They showed a beautiful crisp, clear view of a galaxy tens of millions of light-years away and visible were faint structures as small as 30 light-years across. Hubble had been transformed into the long-awaited telescope that was originally promised.

"Getting those first images was very memorable," Jenkner said, "I think we all breathed a sigh of relief that the servicing mission fixed the flaw of the telescope, and some say it probably saved NASA, too."

SM1 was memorable for Jenkner for another reason: "I had the opportunity to take the woman I was dating at the time to watch the space walks at the Institute. A few months later she married me."

Birthday balloon: A balloon-like bubble created by a super-hot, massive star is appropriately named the Bubble Nebula, and this image was released to celebrate Hubble's 26th birthday. The giant bubble is 7 light-years across — about one-and-a-half times the distance from our Sun to its nearest stellar neighbor, Alpha Centauri. The Bubble Nebula lies 7,100 light-years from Earth in the constellation Cassiopeia. Credit: NASA, ESA and the Hubble Heritage Team (STScI/ AURA)

Hubble would be successfully serviced and repaired in subsequent servicing missions in 1997, 1999, 2002 and 2009, all performed by astronauts to keep the telescope operating with improved instruments and other vital components.

"Every time we had a servicing mission," Cepollina said, "it got more and more complicated, because we wanted the astronauts to do more in less time, and each mission it seemed they had more components to replace on the spacecraft. By the time we got to the second-to-the-last mission, everybody wanted to be a part of it, even those who didn't think this was a good idea in the beginning! It was all exciting and it certainly kept the juices flowing." (You can read about what Cepollina is working on now on page 202.)

"The servicing missions were always nail-biters," said Sembach. "For me, personally, while the missions themselves were great, the buildup to the missions, the discussions, the decisions, the teamwork and the camaraderie built by all the preparation was really exciting."

And, remarkably, as of 2016, the status report on the 26-year-old Hubble reads like a virtually new telescope.

"Hubble is doing great," Sembach said, "and actually in terms of the numbers of scientific papers produced, Hubble had its most productive year ever in 2015, with 846 published papers—more than two every day. The observatory, itself, is in good shape, the instruments are performing beautifully, the subsystems, gyros, reactions wheels, communications, solar panels all look really good. Right now, we are very confident we should be able to have really good science coming from the observatory into the early 2020s."

SCHOOL BUS IN SPACE

Hubble is 43.5 feet (13.2 m) long—about the size of a school bus—and weighs 24,500 pounds (11,110 kg).

One of the most important components of any telescope is its mirror because what makes a telescope powerful is not its ability to magnify objects, but the amount of light it can collect. The bigger the mirror, the more light-gathering power, and, therefore, better vision it has.

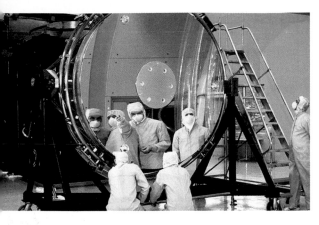

As the Hubble Space Telescope is being built, workers study Hubble's main, eight-foot (2.4-m) mirror in 1990. Credit: NASA

By professional telescope standards, Hubble's mirror isn't that impressive in size. At 8 feet (2.4 m), Hubble's primary mirror is much smaller than the 34-foot (10.4-m) mirror on the Great Canary Telescope on the Canary Islands, currently the world's largest telescope. And Hubble might seem tiny in comparison to the Giant Magellan Telescope, currently under construction in Chile, which will have a light-gathering surface spanning 82 feet (25 m), comprised of seven separate segments that are 27-feet (8.4-m) in diameter.

Hubble makes up for its size with great optics and that all-important placement outside of Earth's atmosphere, giving it remarkable clarity. Hubble can detect objects ten billion times fainter than the unaided eye can detect, with resolution 10 times better than some larger telescopes on Earth, especially in optical and ultraviolet wavelengths. Hubble has proven to be an incredibly resilient and versatile observatory and continues to produce new discoveries and stunning images.

In orbit 343 miles (552 km) above Earth, Hubble circles our planet every 97 minutes, moving about five miles (8 km) per second. When conditions are right, you can actually see Hubble travel across the nighttime sky. You can go to websites like Heavens Above (www. heavens-above.com) to find out when and where satellites like Hubble and the International Space Station will be flying over your backyard.

A NEW LOOK AT THE UNIVERSE

Soon after the astronauts fixed Hubble, the telescope began making observations that changed the face of astronomy. Two early observations brought Hubble into the public eye . . . and into their hearts.

Hubble Space Telescope's image of comet Shoemaker-Levy 9 taken on May 17, 1994, showing a string of 21 icy fragments stretched across 710 thousand miles (1.1 million km) of space, before they hit Jupiter. Credit: NASA, ESA and H. Weaver and E. Smith (STScI)

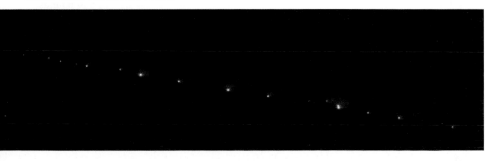

In the summer of 1994, astronomers discovered a comet named Shoemaker-Levy 9 that had come perilously close to Jupiter. The gravity from the giant planet had torn the comet apart into a string of 21 fragments—a string of pearls some scientists called it—and it was apparent the fragments were going to plunge into Jupiter. This is something that had never been witnessed; astronomers were able to use Hubble to get a front-row seat to a spectacular event, and they watched as comet fragments plummeted into Jupiter's cloud tops, exploding like nuclear bombs and creating dark plumes. During press conferences, astronomers ran in with the latest images of the event, creating a buzz of excitement.

One of the most iconic images taken with Hubble is a picture of the Eagle Nebula, a vast cloud of distant gas and dust where stars are being born. Nicknamed the Pillars of Creation, the jaw-dropping photo, taken in 1995, revealed never-before-seen details in star-forming regions. The Hubble image became so popular that it made the cover of *TIME* magazine and has appeared in movies and television shows, been screen-printed on things like T-shirts and pillows and was even on a postage stamp.

From then on, the list of Hubble images reads like a greatest hits compilation: the Horsehead Nebula; the Cone Nebula; planetary views of Saturn, Mars and Jupiter; supernova remnants; and distant galaxies. And the public loved it.

"One of the great foresights came from our first director, astrophysicist Riccardo Giacconi, who from the beginning built a public outreach component into all we do," said Jenkner. "That was perhaps my greatest surprise of the mission—aside from all the scientific findings that changed our view of the universe. That was expected in some degree. What I didn't expect was that we would reach millions of people with our press releases, images and education programs, and that the public would come to think of Hubble as 'their' telescope. It is humbling."

HOW TO CREATE A PRETTY PICTURE

Despite the colorful images the observatory has produced, the cameras on Hubble do not take color pictures. And while Hubble can point very precisely, it's not a point-and-shoot instrument like cameras on Earth.

Hubble Space Telescope image of the cluster Westerlund 2 and its surroundings was released to celebrate Hubble's 25th year in orbit and a quarter of a century of new discoveries, stunning images and outstanding science. Credit: NASA, ESA, the Hubble Heritage Team (STScI/AURA), A. Nota (ESA/STScI) and the Westerlund 2 Science Team

"The fortunate by-product of the science produced by Hubble is that we can make pictures," said Zolt Levay at STScI, who has been producing images from Hubble since just before the first servicing mission. "The color pictures are not really used by science, per se. They really are meant for public consumption, but it's a nice by-product and is a visual representation of what the scientists are studying."

Color images are assembled from separate black-and-white photos taken through color filters. For one image, Hubble has to take three pictures, usually through a red, a green and a blue filter, and then each of those photos gets downlinked to Earth. They are then combined with software into a color image.

While the data from Hubble arrives in muted gray scale, there's a lot of color information embedded in the Hubble data. "It turns out the universe is more colorful than we can imagine," Levay explained. "In fact, there are colors we can't see, light at different wavelengths and energies, and Hubble's cameras are sensitive to that."

Hubble has almost 40 different color filters ranging from ultraviolet (bluer than our eyes can see) through the visible spectrum to infrared (redder than what is visible to humans). This gives the imaging teams infinitely more flexibility and, sometimes, artistic license.

The large Whirlpool Galaxy (left) is known for its sharply defined spiral arms. Their prominence could be the result of the Whirlpool's gravitational tug-of-war with its smaller companion galaxy (right). Credit: NASA, ESA, S. Beckwith (STScI) and The Hubble Heritage Team (STScI/AURA)

"Because the cameras don't work like cameras we use here on Earth, we have to do a little bit of work to get the pictures to look the way they do," Levay said. "This involves a combination of science and art, objectivity and subjectivity. It's no different really from what any photographer might do in printing from a negative or working with raw images from a digital camera."

Levay said he and the imaging team select different colors from the light that comes into the telescope, and they combine them to construct the color images. In general, the colors in the Hubble images are more vivid, and the objects appear much brighter than what our eyes would see. In fact, we couldn't see these objects anyway because they are so distant, faint and, again, sometimes in wavelengths not visible to humans.

But the most important goal of the Hubble imaging team is to produce images that convey as much scientific information as possible. Color is used as a tool to either enhance an object's detail or to visualize what otherwise could not be seen by the human eye, such as ionized gases that make up a nebula or young stars surrounded by dust.

The primary camera on Hubble now is the Wide Field Camera 3 (WFC3), which is basically two cameras: one looks from the ultraviolet through the visible light part of the spectrum, and the other camera looks in the infrared. WFC3 studies dark energy and dark matter and looks at star formation and distant galaxies.

The Advanced Camera for Surveys, a wide-field, extremely sensitive camera, sees in wavelengths from the far ultraviolet to visible light, making it capable of studying some of the earliest activity in the universe. Hubble also has two spectrographs on board. These are instruments that break light into colors and measure the intensity of each color, revealing information about the object emitting the light. The Cosmic Origins Spectrograph works best at studying small sources of light, such as distant stars or quasars, whereas the Space Telescope Imaging Spectrograph can map out larger objects like galaxies or even the locations of black holes.

A 50-light-year-wide view of the central region of the Carina Nebula is one of the most detailed looks ever at this place where both star birth and death are taking place. The fantasy-like landscape of the nebula is sculpted by the action of outflowing winds and scorching ultraviolet radiation from the monster stars that inhabit this inferno. In the process, these stars are shredding the surrounding material that is the last vestige of the giant cloud from which the stars were born. Credit: NASA, ESA, N. Smith (University of California, Berkeley) and The Hubble Heritage Team (STScI/AURA)

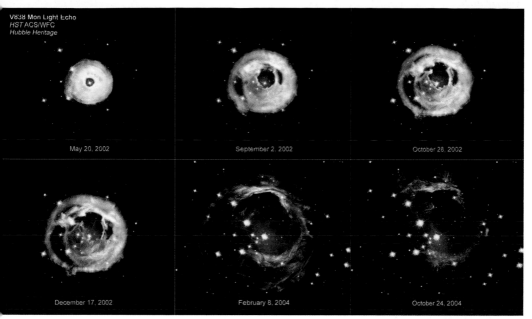

V838 Mon Light Echo
HST ACS/WFC
Hubble Heritage

| May 20, 2002 | September 2, 2002 | October 28, 2002 |
| December 17, 2002 | February 8, 2004 | October 24, 2004 |

An explosion over time. These images show a time sequence of HST images of the light echo around V838 Mon, taken between May 2002 and October 2004. A light echo shows light from a stellar explosion echoing off dust surrounding the star, and these never-before-seen dust patterns were seen after the star suddenly brightened for several weeks in early 2002. The explosion has since dissipated and only the star is now visible. All six pictures were taken with Hubble's Advanced Camera for Surveys. Credit: NASA, ESA and Z Levay

"Astronomy has always been a visual science," Levay said, "where long ago people looked at the sky and watched how things changed. Then Galileo used the telescope, and he was able to see in more detail. Over time, telescopes got more and more sophisticated. Now astronomy is very quantitative, software driven, with statistical analysis. But astronomers still do like to look at pictures, because they are so visually appealing."

Levay's formal training is in astronomy, but he's a self-taught photographer. Framed pictures line the walls of Levay's office at STScI, alternating between Hubble images and his own photography. Levay feels his photography experience is at least as important as the astronomy training he received in regards to creating attractive images from the data.

"It takes a little bit of effort and understanding," he said, "but it's not fundamentally groundbreaking, and the image editing techniques have remained much the same over the years. But we have learned a lot about how to work with the data and the images. I think we have been

able to better appreciate what aesthetic principles contribute to a powerful astronomical image while staying true to the science, and how to communicate the results to the public."

Zolt Levay in his office at the Space Telescope Science Institute. Credit: STScI.

But at the same time, Levay said, the Hubble imaging team does try to create images that are dramatic, interesting and visually compelling, while staying true to the data and the science.

Despite having worked on Hubble imagery for well over 25 years, Levay still seems surprised—astounded, almost—that Hubble imagery has garnered such popular appeal and captured the public's imagination. Levay thinks part of the appeal is that Hubble came along just as the Internet was starting to take off, allowing people easy, instant access to the images.

"None of what I do is rocket science," he said with a smile. "I feel it is pretty straightforward. But what we have going for us is that the material we work with is some of the best astronomical data available! Over the years, the data have been improving and changing with the new cameras installed on HST. This observatory really creates high-quality data, and it's certainly an honor to be able to work with it and do the work I do."

HOW TO GET A PEEK THROUGH HUBBLE'S EYEPIECE

"Hubble has been able to deliver significant progress in practically all areas of astronomy," said Jenkner, as he showed me a book of proposals submitted by astronomers around the world for using Hubble in 2017. "You just need to read the section titles: Extragalactic Programs, Planetary Programs, Galactic Programs, Treasury Programs, Cosmology, the Solar System and so on. Every year we bring in panels of astronomy experts from all over the world to review those proposals."

Unlike other missions that have dedicated science teams who exclusively use the instruments, Hubble is open to any professional astronomer, a concept that many say has contributed to the telescope's incredible productivity. But while every astronomer dreams of using Hubble, only 3,000 observing hours are available each year, and there's a rigorous selection process. The expert panel members rank the proposals according to what they feel offer the best, most challenging science.

"We have no shortage of good proposals," Jenkner said, "and the requests for observing time is greater by a factor of five compared to what we can provide."

As director of the Institute, Sembach reviews the panels' recommendations and makes the final decisions in divvying up Hubble's time, sometimes in parcels as small as a few minutes.

Mystic Mountain: A detailed view of a mountain of dust and gas rising in the Carina Nebula. The top of a three-light-year tall pillar of cool hydrogen is being worn away by the radiation of nearby stars, while stars within the pillar unleash jets of gas that stream from the peaks. Credit: NASA, ESA and M. Livio and the Hubble 20th Anniversary Team (STScI)

"Our schedulers are actually some of our unsung heroes and heroines," said Sembach. "They fit this tremendous puzzle together every week to upload a set of commands to the observatory."

Astronomers who win time on Hubble use special software to submit their observing specifications of what, when and how long they need to observe, and which filters they want to use.

"Our schedulers take those submissions and ingest them into the system, figuring out how to optimize all the different observations these 200-plus programs want to do," Sembach said. "The schedulers produce a year-long, long-range plan, and then every week they build a calendar for the following week, scheduled down to the tenth of a second."

The schedulers take all those observing plans and fold in the time needed for housekeeping tasks like maintaining the orbit or communications—the downlinks of data and command uploads. "That goes like clockwork every week, all year long," Sembach said. "It's really an interesting process."

Each week, the data Hubble sends to Earth could fill over eighteen DVDs. The observations of each astronomer remains proprietary for one year, meaning whoever proposed the observations has one year to process and analyze the data. But after the one-year period is over, any astronomer can download archived data via an Internet database and analyze it from anywhere in the world. Today, half of the papers published with new results come from archival data.

HUBBLE'S LEGACY

The Hubble Space Telescope is named for astronomer Edwin Hubble, who in the 1920s determined that the fuzzy patches of light in the night sky that astronomers originally called **spiral nebulae** were actually other galaxies like our own but very distant. This realization forever changed the view of our place in the universe.

But then in 1929, Edwin Hubble made another astounding discovery: almost all galaxies are moving away from us, and the farther a galaxy is from Earth, the faster it appears to speed away. This notion of an expanding universe formed the basis of the big bang theory, which proposes the universe began with an intense burst of energy at a single moment in time and has been expanding ever since.

This HST photo of the spiral galaxy NGC 3021 was one of several hosts of Type Ia supernovae observed by astronomers to refine the measure of the universe's expansion rate, called the Hubble constant. Hubble made precise measurements of Cepheid variable stars in the galaxy as well, and these stars pulsate at a rate that is matched closely to their intrinsic brightness. This makes them ideal for measuring intergalactic distances. Credit: NASA, ESA and A. Riess (STScI/JHU)

But the rate of expansion—called the **Hubble constant**—was the next question, and part of being able to find the answer depends on knowing the age of the universe. Before the Hubble Space Telescope, astronomers had been able to narrow the universe's age down to somewhere between 10–20 billion years old—not a particularly exact measurement. But with Hubble, astronomers made the determination our universe is approximately 13.7 billion years old, and they hope to narrow it down even more.

"Measuring the age of the universe was one of the things Hubble was designed to do," Sembach said. "Originally, we hoped to determine the age with 10 percent precision then were able to refine it to 3 percent. Now we are trying to do it to 1 percent precision. From a cosmology and astronomy standpoint, that's a remarkable thing to be able to do."

Another finding by the Hubble telescope in 1998 showed that, incredibly, the universe's expansion is actually accelerating. "I think that caught pretty much everyone by surprise," Sembach said, "and is really one of those things discovered with Hubble that really changes your view of the universe."

To explain an accelerating universe, astronomers use the term **dark energy**. While the nature of dark energy remains a mystery, astronomers can clearly see its effects.

"We know dark energy pervades the universe because we've been able to measure the expansion rate of the universe at different times by looking at distant supernovas," Sembach said, "which are dimmer and farther away than expected. Dark energy acts as a repulsive force rather than an attractive force like gravity. It's the equivalent of throwing a ball up in the air and instead of coming back down, it speeds up and keeps going."

The astronomers who determined the accelerating expansion won the Nobel Prize in Physics in 2011. Now the challenge is to determine exactly what dark energy is and how it works, an important mystery to solve since astronomers estimate roughly 68 percent of the universe is made of dark energy.

Gravitational lensing: Abell 370 is one of the very first galaxy clusters where astronomers observed the phenomenon of gravitational lensing, where a foreground galaxy bends and magnifies the light of galaxies located far behind it, distorting their shapes and creating arcs and streaks in the picture, which are the stretched images of background galaxies. Gravitational lensing provides a vital tool for astronomers when measuring the dark matter distribution in massive clusters, since the mass distribution can be reconstructed from its gravitational effects. Credit: NASA, ESA, the Hubble SM4 ERO Team and ST-ECF

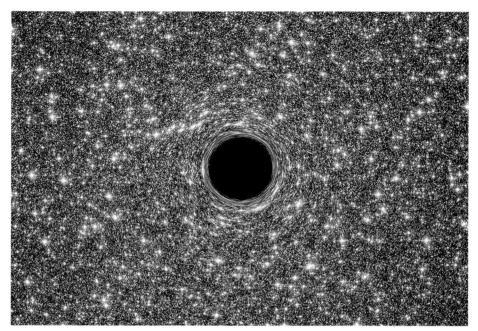

This is an illustration of a supermassive black hole, weighing as much as 21 million suns, located in the middle of the ultradense galaxy named M60-UCD1. The dwarf galaxy is so dense that millions of stars fill the sky as seen by an imaginary visitor. Because no light can escape from the black hole, it appears simply in silhouette against the starry background. The black hole's intense gravitational field warps the light of the background stars to form ring-like images just outside the dark edges of the black hole's event horizon. Combined observations by the Hubble Space Telescope and Gemini North telescope determined the presence of the black hole inside such a small and dense galaxy. Illustration Credit: NASA, ESA and D. Coe and G. Bacon (STScI)

Hubble also was instrumental in exploring another mysterious aspect of the universe, **dark matter**. Observations showed there is far too little visible matter in the universe to make up the remaining 27 percent after dark energy is accounted for, because the stars, planets, dust and gas, or normal matter, make up only about 5 percent of the universe. What is dark matter? It is dark, meaning it is not in the form of material we can see, but it has been inferred for decades because it appears to interact with clusters of galaxies. Dark matter is another mystery to be solved.

Black holes are other objects in space that can't be seen, but their effects are obvious. Even before the Hubble mission, black holes were thought to exist, but again, the space telescope enabled astronomers to survey large numbers of galaxies to show that every galaxy with a bright central stellar bulge indeed contains a supermassive black hole in its center.

"Early in its history, Hubble made essentially the convincing arguments for the existence of black holes almost everywhere," said Jenkner. The day I visited Jenkner at STScI a historic announcement was made by astronomers using the Laser Interferometer Gravitational-Wave Observatory (LIGO) of the existence of gravitational waves, long-sought-after ripples in space-time created by merging black holes. "And so, the culmination of Hubble's determination came earlier today, with the discovery of gravitational waves from two black holes that spun into each other."

Hubble Deep Field: Over ten consecutive days in December 1995, Hubble and the WFPC2 2 stared at a speck of sky no bigger than a grain of sand held at arm's length. In that small patch of sky, more than 1,000 galaxies located billions of light-years away were revealed, each containing billions of stars. Our world and our galaxy suddenly seemed very small. Credit: NASA, ESA, R. Williams (STScI) and the Hubble Deep Field Team

When Hubble launched, only a few **exoplanets**—planets orbiting other stars— were known. Now, Hubble has begun to study these distant worlds.

"Hubble was the first observatory to measure the composition of a planetary atmosphere outside our solar system, an exoplanet atmosphere," Sembach said, "so Hubble opened up that entire field. Before that, no one thought it was possible to make such an observation, but Hubble showed that, in fact, it is possible. It's now observed many such systems, and other telescopes like the Spitzer Space Telescope has also studied exoplanets. We're thinking of future space telescopes that will continue to study exoplanets." (Chapter 5 [page 101]details a dedicated planet-hunting telescope called Kepler.)

Some of the most spectacular displays of Hubble's incredible observing powers came in the Hubble Deep Field studies. Hubble's camera focused on a small area in space that was relatively devoid of any stars, staring at what astronomers thought was an empty area in space. However, some of the deepest images ever taken of our universe are definitely not empty. Every swirl, every smudge, every dot in these images are galaxies, filled with billions of stars. There are over 1,500 galaxies in the original Deep Field photo, which was taken in an area that from our perspective on Earth is the size of a grain of sand held at arm's length. It's mind-boggling to think how many galaxies are actually out there in the rest of the universe, and it shows the incredible size and complexity of our universe.

Hubble has allowed us all to take a journey through time and space, and take an unprecedented voyage to the stars.

ONE MORE TIME

Since the servicing missions had been so successful in keeping Hubble operational, the plan was to continue them. But then the second space shuttle accident occurred in early 2003 as Columbia was returning home from a mission, killing all the astronauts on board.

Again, it was a blow to the entire U.S. space program. And it was just one more hurdle Hubble had to overcome.

After the Columbia accident, then-NASA administrator Sean O'Keefe canceled a scheduled Hubble servicing mission, deeming it too risky. The new protocol for shuttle missions was to make detailed examinations of shuttles in space to look for damage similar to what brought Columbia

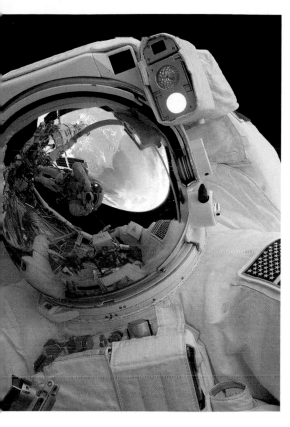

A close-up of Astronaut John Grunsfeld shows the reflection of Astronaut Andrew Feustel, perched on the robotic arm and taking the photo. The pair teamed together on three of the five spacewalks during the final Hubble servicing mission in May 2009. Credit: NASA

down, and if damage was found, the astronauts would use the International Space Station as a safe haven. However, the location of Hubble's orbit around Earth makes it impossible for a space shuttle at Hubble to then reach the International Space Station. This made a mission to Hubble too risky in the estimation of NASA officials at that time. Meanwhile, instruments began failing on Hubble and the outlook for its operational future looked bleak.

"What the NASA administrator and NASA headquarters appeared to underestimate was the reaction of not only the science community but the public," said Jenkner. "After the public outcry there was a rethinking of it, and it turned out to be another Phoenix comeback story for Hubble."

After a reexamination of risks, in October of 2006 new NASA administrator Michael Griffin determined Hubble was such an invaluable tool to astronomers and the public that performing a last servicing mission was an acceptable risk. To mitigate potential dangers to astronauts, a second shuttle stood ready to launch as a rescue mission if anything were to happen.

"The mission was highly successful, with all the drama and human elements that seem to be part of Hubble's great story," Jenkner said, with astronauts installing a new main camera (WFC3) and other instruments designed to help answer profound questions about the origins of the cosmos. They also carried out a number of other maintenance tasks, including repairing the Advanced Camera for Surveys and the Space Telescope Imaging Spectrograph, both of which had quit working.

With the end of the space shuttle program in 2011, no further servicing missions are possible. However, there is hope that one day the iconic telescope might be retrieved or at least deorbited safely.

"One of the things that was put on the telescope during the final servicing mission was a fixture on the back end called a **soft capture mechanism** that will allow a small spacecraft to attach to the telescope to take it out of orbit safely," said Jenkner.

Sombrero Galaxy: A brilliant white core is encircled by thick dust lanes in this spiral galaxy, seen edge-on. The galaxy is 50,000 light-years across and 28 million light years from Earth. This brilliant galaxy was named the Sombrero because of its resemblance to the broad rim and high-topped Mexican hat. X-ray observations suggest that there is material falling into the compact core, where a 1-billion-solar-mass black hole resides. Credit: NASA and The Hubble Heritage Team (STScI/AURA)

HUBBLE AND THE FUTURE

Now, said Jenkner, the goal is to maximize the science return from Hubble in the time still left with the venerable observatory.

"In a sense, Hubble revolutionized astronomy, and it will just be necessary to build on it to continue that path," he said. "So from that point of view, the job is never done and the main reason for the Institute's existence is to never stand still, in a sense, to try to entice the scientists to do bigger and better things, as well as provide them a finely honed tool to do it."

Jenkner explained how they still work to improve the calibrations and accuracy of the instruments on Hubble, even ones like STIS, installed in 1997.

"That's the real challenge and also the beauty of it," he said, "to really wring the last bit of science out of our instruments."

Reiterating the theme that Hubble is better for all its delays, Sembach said when it looked as though the final servicing mission was going to be canceled, Hubble engineers proactively looked at how they could extend the mission.

"We knew gyros would be a limiting factor so we started working on a reduced gyro mode to extend their life," he said. "As it turned out, we did need that reduced gyro mode, and now gyros aren't a limiting factor for Hubble because we now know how to use the gyro resources in a new way. That added a longer life to the mission we didn't think we would have."

Also as a result of the servicing mission delay, Sembach said the infrared detectors in WFC3 were improved and are far better than if the mission had launched as originally scheduled.

"Additionally, we learned how to schedule the observatory more efficiently, so even after seventeen to eighteen years, we were still learning how to do things as efficiently as possible," he said. "The thought that Hubble might only last a few more years forced us to be very efficient, as we wanted to eke out every bit of science we could. After the servicing mission, we went back to that efficient mode and it really helped. So, in retrospect, the delay in the servicing mission had a very positive effect on Hubble, making it a much better instrument."

Engineers continue to assess Hubble's systems to help it remain scientifically viable, conducting other life-extension initiatives to prolong Hubble's operations as long as possible.

I asked Jenkner, who has been at STScI for the duration of the mission, about his favorite memories.

"Just coming to work every day is the best part," he said thoughtfully. "The people here have made my work extremely rewarding."

A colorful collection of 100,000 stars are displayed in this small region inside the Omega Centauri globular cluster, a dense group of nearly 10 million stars. Omega Centauri is one of the biggest star clusters in the Milky Way. Credit: NASA, ESA and the Hubble SM4 ERO Team

There's just one downside to such an extended career. "Those people who you work with, who you consider your friends," he said, "they are not allowed to die. That has provided several difficult times over the life of the Institute."

While the loss of any colleague is difficult, Jenkner said the death of Rodger Doxsey in 2009 was especially trying. Doxsey was the head of STScI's mission office for Hubble and was responsible for the day-to-day operations of the telescope. Many described him as the heart and soul of Hubble, and he made major contributions to his field, so his loss was felt throughout the astronomy community.

Returning to other favorite memories, Jenkner said that attending the space shuttle launch for the final servicing mission was a highlight. "Now that was impressive!" he said. "But, the problem again is the people. By the time they launched, we got to know them so well, so seeing those seven astronauts sitting on top of a firecracker with god knows how many million parts needing to work just right, that makes you choke up, there's no question about that!"

As STScI looks to extend Hubble as long as possible, it soon will be welcoming another mission, the long-awaited James Webb Space Telescope, which the Institute will also oversee. JWST is a powerful tennis court–size infrared telescope that should be able to look back to when the first stars, solar systems and galaxies were formed, providing answers to long-held questions about how the universe came together into what we see today. (Read more about JWST on page 211.)

Jenkner said if Hubble keeps operating into the early 2020s, that would provide a few years of overlap with JWST, and the two of them operating together would be more powerful than either of them operating alone.

What will Hubble's legacy be?

"With Hubble we have learned to never stop asking questions," Jenkner said. "We just need to keep the scientific curiosity alive, and that goes for astronomy, education and life in general."

CHAPTER 4

TRAVELING BETWEEN TWO WORLDS: DAWN

DUALITY, DICHOTOMY AND DAWN

Look up on a clear night and if you're lucky enough to see a meteor streak through the sky, there's a good chance you're seeing a small piece of Vesta, the second-largest object in the asteroid belt.

"There are only three solar system bodies to which we've linked specific meteorites that have fallen to Earth: the Moon, Mars and Vesta," said Marc Rayman, the chief engineer and mission director for a spacecraft named Dawn. "Everyone has heard of the Moon and Mars, but not as many people have heard of Vesta."

Vesta and the dwarf planet Ceres are the two destinations and targets of study for Dawn. The mission was so named because exploring these two ancient worlds is providing information about the very beginnings—the dawn—of our solar system.

"Dawn is helping us understand what the conditions were when Vesta and Ceres were originally made," Rayman said. "It is giving us more pieces in the grand puzzle of how our entire solar system formed and evolved—and perhaps how other planetary systems form around distant stars as well."

These two objects are extremely far away from us: Vesta averages about 219 million miles (353 million km) from Earth, and Ceres is about 257 million miles (414 million km) away. At their discoveries over two centuries ago, these two remote worlds were just "mysterious, faint smudges of light amidst the stars," Rayman said. And prior to the Dawn mission, all we knew about Vesta and Ceres came from ground-based telescopes and Earth-orbiting spacecraft like the Hubble Space Telescope. But even pictures of them from the sharp-eyed Hubble—especially of Ceres—are pixelated and indistinct, leaving more questions than answers.

This artist concept shows NASA's Dawn spacecraft above dwarf planet Ceres, as seen in images from the mission. Credit: NASA and JPL-Caltech

Ceres • January 23, 2004
Hubble Space Telescope • ACS/HRC

Vesta • May 14, 2007
HST WFPC2

Hubble Space Telescope images of Vesta and Ceres, which helped astronomers plan for the Dawn spacecraft's tour of these objects in the asteroid belt. Credit: NASA/ESA, J. Parker (Southwest Research Institute) and L. McFadden (University of Maryland)

Now, the Dawn spacecraft is unraveling the mysteries of these worlds, answering queries such as why Vesta and Ceres seem to be so different from each other even though they are both in the asteroid belt. What accounts for Vesta's strange flattened shape? And what are the bright spots seen on Ceres that seem to be "cosmic beacons, like interplanetary lighthouses drawing us forth," as Rayman described them? Are these areas evidence of bright ice or water, intriguing mineral deposits or—as some have suggested—lights from an alien city?

Unique among all other current robotic planetary missions, Dawn uses a revolutionary **ion propulsion** system. This allows Dawn to do something that has never been done before.

"In nearly six decades of space exploration, Dawn is the only spacecraft to ever orbit two extraterrestrial destinations," said Rayman. "I like to think of it as the first true interplanetary spaceship."

This nearly decade-long mission has overcome challenges and hurdles to complete its mission. And since Dawn has now traveled between two worlds, so much about this mission reflects a definite dualism. The advanced spacecraft itself seems to hover between the realms of science and science fiction, and some of the people who work on the mission fully appreciate—and sometimes embody—those two realms. And while Vesta and Ceres both were categorized as asteroids at various times in history, they represent two very different types of planetary bodies.

"Dawn is truly a historic mission," Rayman said, "studying two fossils from the very beginning of our solar system, and telling us part of the story of our own beginnings."

THE DAWN SPACECRAFT

A spaceship enters orbit around a planetary body, does a little reconnaissance—perhaps you might call it exploring a strange new world. Then the ship departs and heads toward another world, slips into orbit again and conducts more exploration. And the spaceship—flying on a column of blue-green light—would boldly go where none had gone before.

With just the right soaring music in the background, this could be a scene from the classic television series *Star Trek*. Instead, it describes the real adventures of NASA's Dawn spacecraft.

"It is only because of the futuristic, sci-fi-like technology of the ion engines that we can orbit multiple destinations," Rayman said. "We are essentially getting two missions in one."

Bright spots from a distance: This image was taken by Dawn as the spacecraft approached the dwarf planet Ceres on February 19, 2016 from a distance of nearly 29,000 miles (46,000 km). It shows that the brightest spot on Ceres has a dimmer companion. Credit: NASA, JPL-Caltech, UCLA, MPS, DLR and IDA

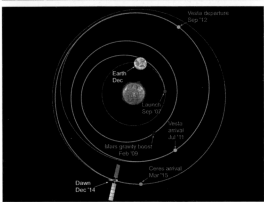

Artist's concept of Dawn spacecraft leaving Earth. Credit: NASA and JPL

This graphic shows the trek of the Dawn spacecraft from its launch in 2007 through its arrival at the dwarf planet Ceres in early 2015. Its journey involved a gravity assist at Mars and a nearly 14-month-long visit to Vesta. Credit: NASA and JPL-Caltech

Dawn launched in September 2007, cruised to Vesta—getting a gravity assist from Mars along the way—and then entered Vesta's orbit in July 2011. The spacecraft spent fourteen months of investigation there before departing for Ceres. It entered Ceres's orbit in early March 2015.

The main goal of Dawn is to gather information about Vesta's and Ceres's composition, internal structure, density and shape. Combined, this data has been helping scientists better understand the processes and conditions present during the early days of the solar system, as well as determine the roles that water content and size play in planetary evolution.

To carry out its scientific mission, Dawn carries a suite of three science instruments: a visible camera, a visible and infrared mapping spectrometer, and a gamma ray and neutron spectrometer. Dawn creates visual, topographic and mineral maps, along with maps of magnetic and gravitational fields, providing a full picture of the surfaces of Vesta and Ceres. Additionally, radio and optical navigation data provide information about the internal makeup of the two bodies.

At the time it launched, Dawn was NASA's largest interplanetary spacecraft from wingtip to wingtip. The spacecraft alone is seven feet nine inches (2.36 m) long—about as long as a large motorcycle. With its solar arrays extended, Dawn is about as long as a tractor-trailer at 65 feet (19.7 m).

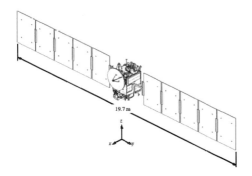

Graphic of the Dawn spacecraft and its instruments.
Credit: NASA and JPL

Marc Rayman explains the representation of the Dawn
spacecraft and its solar arrays in the hallway outside his
office. Credit: Nancy Atkinson

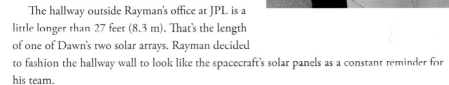

The hallway outside Rayman's office at JPL is a
little longer than 27 feet (8.3 m). That's the length
of one of Dawn's two solar arrays. Rayman decided
to fashion the hallway wall to look like the spacecraft's solar panels as a constant reminder for
his team.

"Since Dawn launched quite some time ago and is now so far away, it's easy to forget how big
the spacecraft really is," he said. "This representation helps us keep in mind that Dawn is a real,
physical thing, not just something at the other end of a computer link or a source of data."

A BRIEF HISTORY OF THE SOLAR SYSTEM VIA VESTA AND CERES

Our solar system came together from a collection of gas and dust surrounding our newly forming
Sun 4.6 billion years ago. Much of the gas and dust in this so-called protoplanetary disk coalesced
to form the planets. Early on, the materials in the disk varied with their distance from the Sun,
with rocky bodies forming closer to the Sun's warmth, and icy bodies forming farther away.

Also in the disk were rocky bodies that had never grown large enough to become planets, and
scientists theorize that large collisions in the early, chaotic solar system pulverized these bodies into
even smaller pieces. These rocky "leftovers" are now the asteroids that travel about our solar system.

Since these remnants contain clues about the early days of our solar system, scientists have
been eager to study them more closely.

The majority of asteroids reside in the asteroid belt, a large doughnut-shaped ring located
between the orbits of Mars and Jupiter, and orbit approximately 186 million to 370 million miles
(300 million to 600 million km) from the Sun.

While Vesta and Ceres are the two biggest objects in the asteroid belt, don't call them asteroids,
Rayman said.

"Many people still mistakenly think of Ceres and Vesta as asteroids," he said. "They really are
not asteroids in any meaningful geophysical sense. These are big places and they exhibit many of
the geological processes characteristic of planetary bodies."

Rayman prefers the term *protoplanets*—meaning they never got quite big enough to be
full-fledged planets.

Vesta and Ceres Size in Context

Lutetia

Vesta

Ceres

Pluto

Unlike most asteroids, Vesta and Ceres are closer in size to Pluto and the Earth's moon

Vesta - 326 miles (525 km) diameter

Ceres - 590 miles (950 km) diameter

Earth's moon

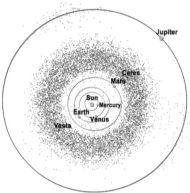

Top: Artist's concept of a very young star encircled by a disk of gas and dust. Credit: NASA and JPL-Caltech

Bottom Left: Ceres and Vesta more closely reflect half-formed planets than space rocks like asteroids. Credit: NASA and JPL

Bottom Right: Artist's graphic of the asteroid belt. Credit: NASA and McREL

One of the biggest mysteries of the two protoplanets is why they are so different from each other.

"While both provide a glimpse into the conditions and processes early in the formation of the solar system, they developed into two different kinds of bodies," Rayman said. "Vesta is a dry, but fascinating rocky world. Ceres, by contrast, may have a large amount of ice and possibly even a subsurface ocean."

The profound differences in geology between these two protoplanets that evolved so close to each other form a bridge, Rayman said, from the rocky bodies of the inner solar system to the icy bodies of the outer solar system.

COMET, ASTEROID OR PLANET?

The current ongoing debate on Pluto's planetary status pales in comparison to what Ceres and Vesta have been through. You'd forgive them if they have an identity crisis.

When Ceres was discovered in 1801, Italian astronomer Giuseppe Piazzi thought he had spied a comet orbiting between Mars and Jupiter. But other astronomers determined it was too big to be a comet, and so Ceres was dubbed a planet—perhaps the "missing" planet that many astronomers theorized should be in the vast space between the rocky red planet and the gas giant Jupiter. But soon, more objects were found with orbits similar to Ceres, marking the discovery of our solar system's main belt of asteroids.

Ceres then turned into an asteroid, laying claim as the largest one in our solar system. But in 2006, the IAU reclassified Ceres as a dwarf planet because of its heft. It has a diameter of about 600 miles (960 km).

Vesta, discovered in 1807, was considered to be a planet, too, but was later reclassified an asteroid with Ceres. It remains the brightest object in the asteroid belt, the only one you can see with your naked eye. But Vesta seems to straddle the line between asteroid and small solar system body (SSSB)—a new term also coined in 2006 by the IAU to describe objects in the solar system that are neither planets nor dwarf planets. Vesta is much bigger than other objects in the asteroid belt: its neighbors are almost all 60 miles (100 km) in diameter or less, while Vesta's average diameter is 320 miles (520 km).

"Vesta itself contains around 8 to 10 percent of the mass of the main asteroid belt," Rayman said, "and Ceres contains about 30 percent. So, Dawn is single-handedly exploring about 40 percent of the total mass of the main asteroid belt."

Remarkable, considering millions of objects orbit the Sun between Mars and Jupiter. However, contrary to popular imagery seen in science-fiction films, the average distance between objects in the asteroid belt is between 621,400 and 1,864,000 miles (1 and 3 million km). The asteroids are spread over such a large volume of space that for the most part, the asteroid belt seems empty. Dawn wasn't dodging other asteroids as it made its travels.

THE MANY WORLDS OF MARC RAYMAN

Vintage toy spaceships and retro sci-fi rockets line Rayman's desk and credenza in his office at JPL, alongside models of real spacecraft, such as NASA's Saturn V and, of course, Dawn. These models are a testament to Rayman's longtime love of both space exploration and science fiction.

"We're exploring two alien worlds using a propulsion system I first heard about on *Star Trek*!" Rayman said with a grin. "What could be more exciting than that?"

At age four, he fell in love with space and knew by the fourth grade he wanted to earn a doctorate in physics (which he did, several years later.) When he was nine, he started writing to NASA and other space and science organizations, and in reply usually received packets of printed materials. Over the years, Rayman has amassed a large collection of space information and memorabilia, with files and collectibles from the space activities of over 50 countries.

Artist concept of the Deep Space 1 spacecraft at asteroid 9969 Braille. Credit: NASA/JPL

Truly a man for all seasons, Rayman is both a scientist and an engineer. In his spare time, he studies particle physics and cosmology. He's a seasoned amateur astronomer, photographer and an avid outdoorsman. He goes dancing every Friday night.

His effervescent personality, along with his well-considered and witty explanations for complex topics, alternates between serious space evangelism and cosmic humor. A prolific writer, Rayman pens a detailed blog about the mission called the Dawn Journal, and he personally replies to questions from readers. He also collaborates on a popular syndicated comic strip *Brewster Rockit: Space Guy!*

"I've never lost my passion for space exploration and the grandeur of scientific discovery," he said. "Working at JPL is a dream come true."

At JPL since 1986, Rayman has been involved in a variety of missions and has helped design others from space telescopes to detect exoplanets to a proposed sample return mission from Mars. Most notably, one mission tested a dozen innovative advanced technologies, all on one spacecraft. Deep Space 1 launched in 1998 and flew by an asteroid and a comet, providing our first-ever close-up views of the nucleus of a comet.

But Deep Space 1 also successfully tested a propulsion system straight out of science fiction called the **ion engine**. This propulsion system had never been used to propel a spacecraft before, and without it, a complex mission like Dawn—and the ability to orbit two different destinations—would never have been possible.

"Ion propulsion is just so elegant," Rayman said. "To be honest, for a lifelong space enthusiast, it's just wonderful to be exploring the solar system with an ion-propelled spacecraft."

ION ENGINES AND THE CONNECTION BETWEEN SCIENCE AND SCIENCE FICTION

Ion propulsion has long been a staple of science-fiction novels, TV shows and movies. All sci-fi fans know that if you want to take a quick sub-light interplanetary trip, use an ion-propelled ship. As Rayman mentioned, it was featured in *Star Trek* (in the "Spock's Brain" episode), and it plays a key role in the *Star Wars* films, as the Imperial TIE fighters are propelled by ion engines (*TIE* stands for twin ion engine).

On the day the movie *Star Wars: The Force Awakens* was released in 2015, Keri Bean, a mission operations engineer for Dawn—and also a full-fledged science-fiction fan—came to work at JPL dressed as Rey, the film's main character. She gave a detailed presentation to her coworkers comparing Dawn to the TIE fighters.

"This idea originally came from Marc Rayman, but I made a full chart comparing the onboard computers and navigation systems," Bean said. "Both spacecraft have two solar array panels, and since Dawn has three ion engines compared to the TIE's two, Dawn is actually a TRI fighter."

Left: Artist's comparison of the Dawn spacecraft, which has three ion engines, and a Star Wars TIE (Two Ion Engine) Fighter. Credit: NASA/JPL

Right: Keri Bean dressed as Rey from Star Wars. *Credit: Keri Bean*

Fully into the research, Bean discovered that an early proposal for Dawn included a laser altimeter, which would have compared nicely to the TIE fighters' green laser cannons, but, alas, it was not included in the final plans. "That's too bad," she said, "because then Dawn would have been a true TRI fighter."

Part of the fun of working on the Dawn mission, Bean said, is how science and science fiction seem to tie together.

"Ion engines came out in science and sci-fi at about the same time," she said, "and they've sort of coexisted ever since. Usually science derives from sci-fi or the other way around, so it is interesting that writers and scientists came up with this idea independently at about the same time."

American rocket pioneer Robert Goddard discussed ion propulsion in his writings in the early 1900s. About the same time, UK writer and astronomer Donald W. Horner published an early space fiction story titled *By Aeroplane to the Sun*, which included an interplanetary voyage using ionic propulsion.

It wasn't until 1959 that NASA tested an early version of ion engines in a vacuum chamber, and around that same time the popular comic strip *Dick Tracy* featured the space coupe, a noiseless electric spaceship that enabled easy travel between Earth and the Moon. The next year, two ion engines were tested in a short suborbital flight on the SERT-1 satellite; only one worked.

Development of ion propulsion was mostly shelved during NASA's push to send astronauts to the Moon, but about the time the first *Star Wars* movie was released, interest resumed. In the 1990s, JPL started the NASA Solar Electric Propulsion Technology Applications Readiness (NSTAR) project with the purpose of developing ion engines for deep-space missions.

From 1996 to 1997, a prototype ion engine at JPL began a long-duration test in a vacuum chamber simulating conditions of outer space. The engine magnificently logged more than 8,000 hours of operation, and with that accomplishment, NASA wanted to test ion propulsion in space. Deep Space 1 flew its mission with greater success than anyone could have imagined, thrusting for 16,000 hours. Later, ESA flew the ion-powered SMART-1 to orbit the Moon from 2003 to 2006.

But the real connection between science and science fiction comes from the name itself. Thanks to Rayman, it's called ion propulsion.

Left: The NSTAR ion engine. Credit: NASA Glenn Research Center

Right: The NSTAR ion thruster for the Deep Space 1 spacecraft during a hot fire test at the Jet Propulsion Laboratory. Credit: NASA Glenn Research Center

Opposite Page: The Apollo 11 Saturn V rocket launched on July 16, 1969, from Kennedy Space Center, carrying the first astronauts to land on the Moon. Credit: NASA

At the time the Deep Space 1 mission was being readied, the type of propulsion it would use didn't have an official name yet. Some wanted to call it NSTAR, others wanted to call it SEP for solar electric propulsion. Rayman had another idea.

"Rather than use a name most technical people were unfamiliar with, it seemed more reasonable to use one that other engineers would understand right away," he said. "That would help when it came time for subsequent mission designers who might consider using the technology."

But as a lifelong space enthusiast, Rayman also had an additional consideration. "NASA uses taxpayers' money, and in my view, we do our work not only for the benefit of scientists and engineers, but also for those taxpayers," he said. "I have been a strong advocate of helping the public be engaged in what we do. And to me, part of our responsibility is to translate difficult concepts into terms that people can understand."

Eager to find ways to capture the public's interest, Rayman chose to call it ion propulsion.

"That is a fully accurate description," he said, "and although I have been a fan of science longer than science fiction, I do very much enjoy science fiction, and I knew about the references to ion propulsion in *Star Trek*, *Star Wars* and other places. And even if you don't know what ion propulsion really is, it sounds a lot more intriguing."

And that's the little-known tale of how this remarkable propulsion system got its name.

CONVENTIONAL ROCKETS VS ION PROPULSION

The basic concept for all space propulsion systems is the same: create enough thrust to propel a spacecraft. In simplest form, a rocket has an enclosed chamber that puts a propellant—usually a gas or a liquid—under pressure, and when the fuel is pushed out the back of a rocket, it provides a thrust that boosts the spacecraft in the opposite direction. Think of a balloon filled with air: when you release the nozzle, the air (a gas) escapes and the balloon zips around the room.

When most people think of rockets, conventional chemical rockets likely come to mind, such as the big, powerful ones used to launch the Apollo missions to the Moon or the space shuttle to orbit. These rockets produce thrust from a chemical reaction that heats the propellant, which is then channeled out of the rocket at high speed in the form of a powerful flame. This provides a huge amount of thrust over a short period of time to push the spacecraft off Earth and into space.

An ion engine, however, uses a different technique: it produces a small amount of thrust over a much longer time. And that time makes a big difference.

Ion engines use a gas, usually xenon, a nontoxic noble gas (a gas that occurs naturally). The gas is ionized, which means it gets zapped with an electrical charge, making the particles repel each other. The ionized gas is accelerated through an electric field at the back of the engine, producing thrust.

While conventional rockets thrust for a few minutes before they consume all their fuel and then coast to their destination, ion engines are almost constantly active, using a small amount of fuel.

"The reason ion propulsion is so much more efficient than conventional chemical propulsion is that it can turn electrical energy from solar panels into thrust," Rayman explained. "Chemical propulsion systems are limited to the energy stored in the propellants."

Dawn carries 937 pounds (425 kg) of xenon propellant, and uses only about 3.25 mg of xenon per second (about 10 ounces over 24 hours) at maximum thrust. In comparison, the rockets used to get the space shuttle into orbit consumed 2.4 million pounds (1.1 million kg) of fuel in only 100 seconds.

There's one downside to ion engines: they aren't powerful enough over a short amount of time to lift a spacecraft off Earth's surface. For that, you need the sudden, swift acceleration that only chemical rockets can provide to overcome the pull of our planet's gravity.

Rayman calls ion propulsion "acceleration with patience." No warp speed here, it takes Dawn about four days to go from 0 to 60 miles per hour (96 km per hour)

Yep, you read that right.

Rayman held up a piece of paper. "The ion engine pushes on the spacecraft as hard as a single piece of paper pushes on my hand," he said.

As small and insignificant as that push might seem, Rayman explained, over the course of the mission the total change in velocity from the ion propulsion is comparable to the push provided by the big Delta II rocket that lifted Dawn from Earth into space and beyond Earth's orbit. This is because the ion propulsion system has operated for thousands of days, instead of the minutes during which the Delta rocket burned.

Once in space, the ion engine works like a breeze.

Dawn launched at dawn (7:34 a.m. EDT) from Cape Canaveral Air Force Station, Sep. 27, 2007. It's mission is to learn about the dawn of the solar system by studying Vesta and Ceres. Credit: KSC/NASA

"In the zero-gravity, frictionless conditions of spaceflight, gradually the spacecraft builds up speed," Rayman said. "Dawn's total velocity change since its launch amounts to 24,600 miles per hour (39,600 km per hour). You can compare that to the 17,500 miles per hour (28,160 km per hour) it takes to go from the surface of Earth to low Earth orbit."

Ion propulsion is uber-efficient with fuel during its in-flight maneuvers. "A typical Mars orbiter could consume more than 600 pounds (272 kg) of propellants to enter orbit around the red planet," says Rayman. "With its ion propulsion system, Dawn could do it with less than 60 pounds (27 kg) of xenon."

For a conventional rocket to do the type of maneuvering Dawn can do—coming and going between two worlds—it would have to carry so much fuel that it would be too heavy to launch from Earth.

Dawn has three twelve-inch (30-cm)-diameter ion propulsion engines; however, only one thruster operates at a time. For the entire mission, total thrust time will be about 2,020 days—nearly six years. The only interruptions amount to a few hours each week so the spacecraft can turn to point its antenna to Earth to send and receive data.

Dawn also uses small hydrazine thrusters and devices called **reaction wheels** for controlling the spacecraft orientation, or attitude. While the ion engines have worked nearly flawlessly over the entire mission, the reaction wheels have provided one of the biggest challenges for the Dawn engineering team. What occurred in 2012 could have disastrously ended the mission.

ENGINEERS SAVE THE DAY

Bean says her job as a Dawn mission operations engineer looks like a normal desk job, just like millions of other people's.

"I sit in front of my computer, I answer emails, I have meetings," she said. "But the end result is probably a little different."

Bean tells a spacecraft what to do. And it does it.

Her specialty with the Dawn mission is called **science planning**. "I like to think of it as a translator," she said, "where I'm translating between the science team and the engineering team, making sure we get all the data the scientists want while keeping it within the engineering constraints."

While the Curiosity rover team has hundreds of scientists and engineers, Dawn has about 40 engineers and 80 scientists from all around the world. "We're a tiny team, but we're getting a massive amount of interesting science back," Bean said.

The Goldstone Deep Space Communications Complex, located in the Mojave Desert in California, is one of three complexes that comprise NASA's Deep Space Network (DSN). The DSN provides radio communications for all of NASA's interplanetary spacecraft and is also utilized for radio astronomy and radar observations of the solar system and the universe. Credit: NASA/JPL

And unlike the Curiosity mission that communicates with Earth almost every day, Dawn links up with the DSN only once or twice a week. Therefore, all communications—whether Dawn is sending data back to Earth or receiving new instructions—have to be planned out well in advance.

Bean joined the mission in 2013 but is honored to be part of the team of deft and resourceful Dawn engineers who have literally saved the mission. They had to develop innovative ways to fly the spacecraft due to unforeseen—and potentially catastrophic—events. Their ingenious solutions would make a crackerjack engineer like Scotty on *Star Trek* proud.

Dawn's reaction wheels are gyroscope-like devices used to both stabilize and turn the spacecraft to point its instruments toward their target or to point the main antenna toward Earth. There are four reaction wheels on board, and under normal circumstances three are required. However, Dawn experienced the failure of two reaction wheels, one in June of 2010, the other in August of 2012.

"Two failures like that could truly be dire," Rayman said. "Other missions have essentially had catastrophic consequences due to reaction wheel failures. It really is a testament to the Dawn flight team that not only did we overcome it, but we have now accomplished and even exceeded all of the original objectives of the mission."

Initially, the flight team devised a work-around using the hydrazine thrusters to compensate for the failed reaction wheels, and using the two remaining wheels only sparingly. But since the hydrazine is limited and extremely critical—whenever the hydrazine runs out, the mission ends—the engineers undertook an aggressive campaign to conserve this fuel. They analyzed more than 50 different options, and now instead of using the projected 27.6 pounds (12.5 kg) of hydrazine during the cruise and orbit insertion at Ceres, Dawn used just 9.7 pounds (4.4 kg), an astonishing 65 percent reduction.

"Thanks to their remarkable ingenuity and resourcefulness, the team has also devised a detailed plan that should allow Dawn to complete its extraordinary mission using only the hydrazine thrusters if the other two reaction wheels also fail," Rayman said.

Dawn launched with 101 pounds (45.6 kg) of hydrazine, and when it left Vesta, it still had 71.2 pounds (32.3 kg). With the two remaining wheels currently operating, they are providing a bonus reduction in hydrazine use, effectively extending the mission.

"To be able to recover from an unplanned and very serious failure like this is quite a testament to our engineers' creativity," Rayman said. "And now, despite the tight constraints of flying the spacecraft differently, the team has been able to add bonus objectives to the mission. That's amazing."

DAWN AT VESTA

Remarkably, more meteorites collected on Earth come from Vesta than from the Moon or Mars. Even accounting for the 842 pounds (382 kg) of samples that Apollo astronauts brought back from the Moon, there are far more samples from Vesta, well over 1,000 pounds (454 kg) of material.

Left: Three slices of a class of meteorites that NASA's Dawn mission has confirmed as originating from Vesta. The meteorites, known as howardite, eucrite and diogenite meteorites, were viewed through a polarizing microscope, where different minerals appear in different colors. Credit: NASA/University of Tennessee

Right: Dawn's framing camera obtained this image, showing the south pole of Vesta and the Rheasilvia Impact Basin. Credit: NASA/JPL-Caltech/UCLA/MPS/DLR/IDA

 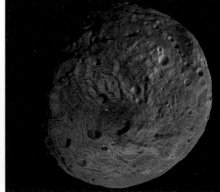

An image taken by NASA's Dawn spacecraft on July 24, 2011, shows chasms along the equator of the asteroid Vesta, including Divalia Fossa, which is larger than the Grand Canyon. The chasms were likely created from impacts to Vesta's south pole region. Credit: NASA/JPL-Caltech/UCLA/MPS/DLR/IDA

"Six percent of all meteorites that fall to Earth—that's 1 in 16—come from Vesta," said Rayman. "That's a remarkable number. We had a good idea that they were pieces of Vesta because of spectroscopy studies done from Earth since the 1970s, but now Dawn has clinched the story, proving these meteorites indeed originated from Vesta."

Spectroscopy splits the light of a distant object into its component colors, which tells scientists what the object is made of. The spectrum of the meteorites and Vesta matched.

Now, detailed spectral data from Dawn has not only confirmed these meteorites came from Vesta, but also determined that—amazingly—all these meteorites came from one giant impact region at Vesta.

Dawn photographed that impact site, named Rheasilvia, a massive crater more than 300 miles (500 km) in diameter. Inside the crater sits a massive mountain, 110 miles (177 km) across at the base with a peak that soars two-and-a-half times higher than Mount Everest, the highest mountain on Earth.

Rheasilvia actually intersects with an older impact site called Veneneia, which is 250 miles (402 km) across. With these two intersecting craters, the entire southern hemisphere is deformed, giving Vesta the look of a flattened gray basketball.

"The giant impacts in Vesta's southern hemisphere created so much energy that the whole body reverberated and almost broke up," Rayman said. "As the energy went through the interior, it left a network of 90 chasms near the equator, some of which rival the Grand Canyon in size. This was completely unanticipated and is just amazing. We have nothing like this on Earth."

But yet Vesta is also quite Earthlike. Dawn compiled the first map of Vesta, revealing an exotic and diverse geologic landscape. Similar to Earth, Mars, Venus and Mercury, Vesta has ancient basaltic lava flows on the surface and a large iron core. It has tectonic features like Earth: rift valleys, ridges, cliffs and hills.

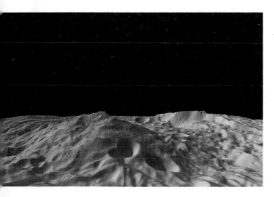

A unique, oblique or sideways view of Vesta's south polar region, created from a shape model. Credit: NASA/JPL-Caltech/UCLA/MPS/DLR/IDA/PSI

A full view of Vesta, and visible is the 'snowman' shaped set of craters. Credit: NASA/JPL-Caltech/UCLA/MPS/DLR/IDA/PSI

Therefore, Dawn confirmed what scientists had theorized: Vesta resembles a small planet more than it does a typical asteroid. It is a surviving "baby" planet from the dawn of the solar system, enduring the intense collisional environment in the asteroid belt. Nothing that telescopes see today in the asteroid belt quite matches what Dawn has seen at Vesta. Imagery shows Vesta is riddled with crazy configurations of craters, including a triple crater shaped like a snowman.

"Life in the main asteroid belt is a tough, rough-and-tumble environment," Rayman said. "Vesta shows us what four million years of being pummeled by rock looks like. But it may be the only remaining example of the original objects that came together to form the rocky planets like Earth and Mars."

Having studied Vesta in detail—and longer than originally planned—Dawn departed Vesta for a two-and-a-half year journey to Ceres.

NO MINUTES OF TERROR

With Dawn conducting unprecedented maneuvers of entering orbit twice and leaving orbit once, you might expect several moments of terror similar to the Curiosity rover's landing. However, when Dawn went into orbit around Vesta, the mission control room was dark and empty. Marc Rayman was out dancing with his wife. Similarly, when Dawn went into orbit at Ceres, most members of the mission team were all at home, probably asleep.

"This was an entirely different process from other orbital missions," said Rayman, "which has nothing to do with Vesta and Ceres, but instead is all about the ion propulsion."

Instead of firing the thrusters full blast, Dawn gently slipped into orbit, letting gravity grab the spacecraft with a light tug as it gradually spiraled in ever-smaller loops until reaching the desired orbit. Similarly, when it left Vesta, Dawn spiraled in increasingly larger loops until the spacecraft was far enough away that Vesta's gravitational grip could no longer hold it.

"When Dawn goes into or out of orbit, it uses the most gentle, graceful, elegant movements in order to accomplish these truly remarkable events," said Rayman. "There are no 'do-or-die' moments. If there had been some problem or anomaly that affected the spacecraft when we were attempting to go into orbit, we would have just gone into orbit another day. That's a capability not available with other types of propulsion."

After Dawn's launch, the team spent years gradually shaping Dawn's orbit around the Sun so that when it got near Vesta four years later, Dawn was already going around the Sun in practically the same path as Vesta's, approaching its target with a relative speed of about 55 miles per hour (88 kmh).

"So, if a problem arose and we hadn't been able to continue the original flight plan, Dawn would be moving at such a slow relative speed that we simply could have adjusted the orbit to arrive later," Rayman said. "That removes that whole problem of only getting one shot at it."

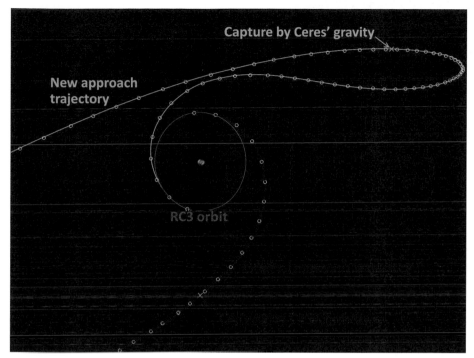

Capture by Ceres' gravity

New approach trajectory

RC3 orbit

Graphic showing the original and the new approach to Ceres for the Dawn spacecraft. Credit: NASA/JPL.

While Rayman admits the do-or-die moments are exciting from a public perspective, he doesn't feel the Dawn team missed out on any excitement.

"To me, the real drama is getting to explore alien worlds," he said.

But there have been some nerve-wracking, nail-biting moments during the mission, such as the reaction wheel failures.

Also, in September 2014 as Dawn was on its way to Ceres, the ion propulsion system suddenly stopped working. Analyzing the spacecraft from Earth, the Dawn team determined a cosmic ray—a particle of space radiation—just happened to hit an electrical component on the spacecraft, shutting down the ion engine and putting the spacecraft in safe mode. The timing couldn't have been worse.

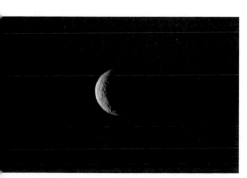

This image of Ceres was taken by the Dawn spacecraft on March 1, 2015, just a few days before the mission achieved orbit around the previously unexplored world. The image shows Ceres as a crescent, mostly in shadow because the spacecraft's trajectory put it on a side of Ceres that faces away from the Sun. Credit: NASA/JPL-Caltech/UCLA/MPS/DLR/IDA

"While we weren't concerned we wouldn't get to Ceres," Rayman said, "we had already developed a very intricate plan of how we would go into orbit around Ceres. Now all of a sudden because of one subatomic particle, we had to create an entirely new trajectory. That was not the most relaxed time of flying this interplanetary spaceship!"

The team worked to diagnose what had occurred, figure out how to correct it, reconfigure the spacecraft for normal operations and—on short notice—develop a new flight path.

"Getting the spacecraft out of safe mode isn't fun," Bean said. "The engineers had to work around the clock. But it later paid off because we ended up with some very interesting science observations and some gorgeous images we originally would never have taken."

The new trajectory had the spacecraft approaching from Ceres's dark side, and as Dawn looked down over the north hemisphere, Ceres appeared as a beautiful crescent, half dark and half light.

"That's a perspective we originally weren't planning to see," Rayman said. "To me, this looping trajectory is one that any hotshot spaceship pilot would be proud to fly! It went flawlessly and also led to some very striking views of Ceres."

And as Dawn grew ever-closer to Ceres, those intriguing bright spots became brighter and more numerous.

"You can't help but be mesmerized by these glowing beacons," Rayman said, "seemingly illuminating the way. It makes you want to send a spacecraft there to find out what they are, and that's exactly what we've done."

DAWN AT CERES

"Ceres is a big place," Rayman said. "It has 37 percent of the area of the continental United States. Just think of how vast, varied and beautiful the geography, topography and geology of our own country is. That illustrates the fascinating variety we are seeing at Ceres."

Dawn has now spotted intriguing landforms and features across Ceres that show it to be a unique world. Of course, the bright areas in a crater called Occator are among the most captivating. Hinted at more than a decade ago by Hubble Space Telescope, the bright regions remained mysterious. They've also been the source of much speculation by the public.

Rayman gives many public presentations and answers questions posed online. He said one of the most often-asked questions is if these bright areas could be lights of an alien city.

The bright central spots near the center of Occator Crater are shown in enhanced color in this view from the Dawn spacecraft. The view was produced by combining the highest resolution images taken in February 2016 at an image scale of 115 feet (35 m) per pixel with color images obtained in September 2015 at a lower resolution added. Credit: NASA/JPL-Caltech/UCLA/MPS/DLR/IDA/PSI

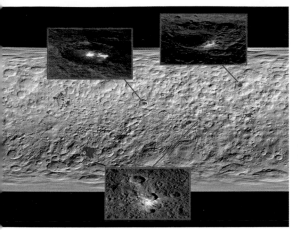

This map of Ceres, made from images taken by NASA's Dawn spacecraft, shows the locations of about 130 bright areas across the dwarf planet's surface, highlighted in blue. Most of these bright areas are associated with craters. Three insets zoom in on a few areas of interest. Occator Crater, containing the brightest area on Ceres, is shown at top left; Oxo Crater, the second-brightest feature on Ceres, is at top right. Credit: NASA/JPL-Caltech/UCLA/MPS/DLR/IDA

Rayman's response? "That's ridiculous! How do we even know the Cereans live in cities? Maybe they live in rural areas; maybe they live in large states. We just don't know enough about the Cereans to know what their geographical distribution is like."

"That's my answer from the perspective as a science-fiction fan," he smiled. "Admittedly, even though these bright regions are surely not evidence of life, they can't help but make you think about that sort of thing. And it reminds you that we live in a vast, complicated, beautiful cosmos and there simply has to be life out there, somewhere . . ."

But Dawn scientists have an idea of what the bright areas might be: large deposits of mineral salts left over from subsurface water that has sublimated away. Details about the deposits reveal they may be similar to sodium carbonate and sodium bicarbonate. On Earth, we commonly call them washing soda and baking soda.

"While everyone's favorite idea about the bright areas has been aliens," said Bean, "I think it is ironic that it might be something so common that you can find it in any grocery store on Earth."

Ceres has more than 130 bright areas, most of them inside impact craters. And the global nature of these bright regions, say Dawn scientists, suggests that Ceres has a subsurface layer that contains briny water ice. The impacts that formed the craters would have unearthed the mixture of ice and salt.

Rayman said these salts are reflective and look bright to our eyes. There also appears to be a diffuse haze of water vapor above Occator Crater and other bright regions, meaning the ice is evaporating today. This might be a confirmation of the detection of water vapor around Ceres in 2014 by ESA's Herschel Space Telescope.

Surprisingly, Ceres is an active body.

Ceres' Oxo Crater (right) is the only place on the dwarf planet where water has been detected on the surface so far. Credit: NASA/JPL-Caltech/UCLA/MPS/DLR/IDA/PSI

"Ceres is the only large body of rock and ice in the inner solar system," Rayman said, "and it doesn't experience any tidal forces to heat it up like the moons of Jupiter and Saturn do. But it does have more sunlight falling on it, so it is really fascinating to be learning the dynamics of this world. This finding is consistent with there being a lot of water there, and it is revealing things about the interior."

Though scientists know with certainty that Ceres has water, the goal is to figure out whether the water is completely frozen or a liquid floating underneath the surface. A few things point to subsurface ice: the density of Ceres is less than that of Earth's crust, and the surface bears spectral evidence of water-bearing minerals. Scientists estimate that if Ceres were composed of 25 percent ice, it may have more water than all the fresh water on Earth.

Dawn is also studying an unusual mountain on Ceres called Ahuna Mons.

"At first, it was called the Lonely Mountain," Rayman said. "It is in an area that is otherwise relatively smooth, and quite abruptly it transitions to a conical mountain. It has bright streaks, is dark on one side and light on the other and looks as though something has run down the steep sides. Regrettably, some people mistakenly called it a pyramid, but we could see it is more like a cone with a flat top."

And Ahuna Mons isn't just a little hill. It rises to about 20,000 feet, the height of Denali, the highest peak in North America.

"Whether it turns out to be a volcano or not," Rayman said, "this feature will likely reveal something important about the interior geological processes at Ceres. This is an active, complex world, not just a hunk of rock and ice. It's a real world!"

TO THE END . . .

Dawn's primary mission was scheduled to end by late 2016, but there was an outside chance it could operate for months longer, until the hydrazine fuel for its maneuvering thruster system ran out. And then, despite the issues with the reaction wheels and an uncertainty of how much hydrazine was available, Dawn engineers proposed an audacious idea: use the ion propulsion to take Dawn out of Ceres' orbit and conduct a flyby of another object in the asteroid belt, asteroid Adeona. However, while NASA decided to extend the Dawn mission, they concluded "long-term monitoring of Ceres has the potential to provide more significant science discoveries than the

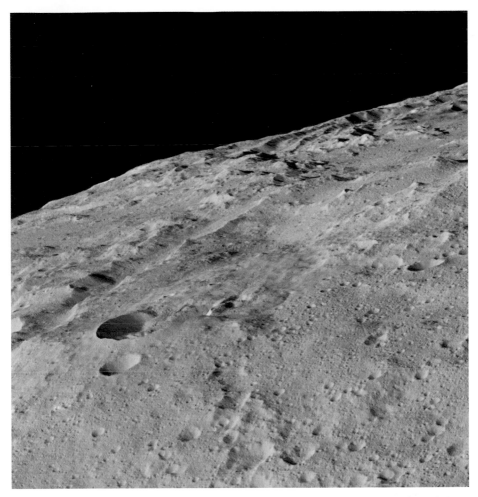

Ceres, taken in Dawn's low-altitude mapping orbit around a crater chain called Gerber Catena. Credit: NASA/ JPL-Caltech/UCLA/MPS/DLR/IDA

Adeona flyby," said Jim Green, NASA Planetary Science Division Director. So, Dawn will stay at Ceres. But it certainly was an exciting proposition and a testament to the benefits of ion propulsion.

Still, the major limiting factor for Dawn's longevity is the hydrazine. Bean said it's hard to estimate the amount of the remaining fuel.

"We use the hydrazine fuel to turn the spacecraft so it can point its antenna toward Earth to communicate, and toward its target so it can make observations," explained Bean. "We'll just keep using it until the fuel runs out and get as much data as we can."

When that fuel is gone, Dawn will remain in a stable orbit around Ceres, perhaps for decades.

"Dawn will become a little moon of Ceres," Bean said. "When the fuel runs out, there will be no chance of communicating with it later, because without fuel it won't be able to control where it is pointing, so it won't be able to keep the solar panels pointed toward the Sun. That means it will run out of power shortly after losing all the fuel."

Artist concept of the Dawn spacecraft in flight. Credit: NASA/JPL

The end of Dawn's mission, therefore, will come quietly and without drama, much like how it arrived at the worlds it explored.

"It will be sad," Bean said wistfully. "It's been a long, unprecedented mission so far, and one day we aren't going to hear from it anymore. And that's going to be it."

But there are plenty of discoveries waiting to be made at Ceres. Rayman reflected on the mission and the importance of exploration.

"While we are confined to the neighborhood of our humble home planet, we reach out with our spacecraft and undertake these grand adventures," he said thoughtfully. "We do this in order to comprehend the majesty of the cosmos and to act upon this passion we feel for exploration. Who hasn't looked at the night sky in wonder? Who hasn't wanted to go over the next horizon and see what's beyond? Who doesn't long to know the universe? Anyone who has ever felt any of those feelings is a part of this mission, and being able to share this experience is what I think is the most exciting, gratifying, rewarding and profound aspect of exploring the cosmos."

CHAPTER 5

HUNTING FOR PLANETS: KEPLER AND THE SEARCH FOR OTHER WORLDS

A DIFFERENT VIEW

On a clear night in 2012, astrophysicist Natalie Batalha went out for a run. A star-filled sky hung above her, but she didn't really notice. Instead, replaying through Batalha's mind were calculations she and her team with the Kepler space telescope had been working on, trying to determine the potential number of planets orbiting other stars in our galaxy.

The results were staggering. The data from Kepler suggested every star in our Milky Way galaxy has at least one planet. With estimates ranging between 100 and 300 billion stars in the Milky Way, that meant there could be hundreds of billions of planets out there. If only a small fraction of these worlds—say 10 percent—were the size of Earth, and if just 10 percent of those Earth-size worlds were the right distance from their star in what is called the **habitable zone**, there could be at least a billion potentially habitable Earth-size worlds in our galaxy.

An artist's concept of the Kepler telescope in space. Credit: NASA/Ames Research Center

Batalha, Co-Investigator of the Kepler Mission, holding an artist's model globe of the first confirmed rocky exoplanet, Kepler 10b. Credit: NASA/ Kepler Mission

There could be planets—and possibly life—everywhere. Less than 30 years ago, astronomers weren't sure of any other planets outside of our own solar system. This remarkable new approximation of the extent of other potentially habitable worlds gave Batalha pause. She stopped running and looked up at the sky.

"I had been thinking about the average number of planets per star in a very academic way for quite some time," Batalha said, nearly four years later from her office at NASA's Ames Research Center in California, where she is Kepler's mission scientist. "But suddenly on that one night when I looked up at the sky, in just a microsecond I saw those pinpoints of lights not only as stars but as planetary systems; other solar systems beyond our own. It was a dramatic shift in the way I observed the universe."

While the statistics from the Kepler mission are certainly a more integral part of Batalha's life than for the general public, her experience symbolizes the shift in humanity's knowledge and understanding of our place in the cosmos. This shift has been taking place over the past several years primarily because of the findings by the Kepler spacecraft. This mission has undoubtedly changed our view of the universe.

The Kepler spacecraft launched from launch pad 17-B at Cape Canaveral Air Force Station in Florida on a Delta II rocket, on March 6, 2009. Credit: NASA, Regina Mitchell-Ryall and Tom Farrar

ARE WE ALONE?

The question "Are we alone?" is centuries old, but conclusive data about exoplanets—planets orbiting other stars—is relatively new.

"For me, the most exciting question you can possibly pose is 'are we alone,'" said Thomas Barclay, a senior research scientist with the Kepler mission. "Being part of this large collaboration of people whose quest it is to answer this question has been extremely exciting and humbling."

Barclay pointed out there are usually two opposing viewpoints on the topic: either the universe is so big that there has to be other life out there somewhere, or Earth is rare and unique, with so many coincidences having to converge for life to spring up on Earth that it might be hard to replicate.

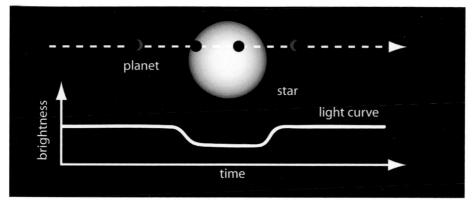

When a planet crosses in front of its star as viewed by an observer, the event is called a transit. Transits can produce a small change in a star's brightness. By measuring the depth of the dip in brightness and knowing the size of the star, scientists can determine the size or radius of the planet. The orbital period of the planet can be determined by measuring the elapsed time between transits. Once the orbital period is known, scientist can determine the average distance of the planet from its star. Credit: NASA Ames

As far back as ancient Greek writings, humans have wondered if other worlds and civilizations might exist. But prior to 1994 was no real data about planets around other stars—just gut feelings and speculation. And scientists don't like to speculate, they want data to find the answers.

"This is a hypothesis you can't easily test," Barclay said, "but the Kepler mission represents the faith that the answer is we are *not* alone, and we are going to do our darnedest to find the other life that is out there."

The Kepler space telescope, launched in 2009, is designed to determine what fraction of the hundreds of billions of stars in our galaxy has planets. In particular, Kepler looks for Earth-size or smaller rocky worlds, and the science team also wants to determine if other solar systems are like ours or vastly different. Because life as we know it requires liquid water, Kepler focuses its efforts looking for exoplanets in or near the habitable zone around a star—also called the Goldilocks zone—where it is not too hot and not too cold for liquid water to exist.

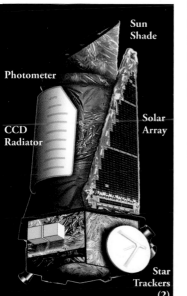

A graphic showing the location of the various parts and the photometer on the Kepler spacecraft. Credit: NASA Ames

For the most part, these worlds are too far away to be seen with telescopes. And Kepler itself can't actually "see" the planets, either. Instead, it searches for exoplanets using a specialized technique called the transit method. When a planet transits, or passes, in front of its star, it blocks a small portion of the star's light, like a mini eclipse. We see the same effect in our solar system when from our vantage point on Earth, Mercury or Venus passes in front of the Sun.

But planets are tiny compared to stars so the amount of light they block is incredibly small, especially since these stars and exoplanets are so far away. However, Kepler can detect even the dimmest of these winks of light, registering changes in brightness of only 20 parts per million. What does that mean? With Kepler watching an incredible 165,000 stars at once, astronomers have compared this feat to viewing a highway full of cars at night from several miles away and being able to detect a flea crossing in front of one headlight. And to confirm the sighting, they need to see the flea at least once more to make sure it wasn't an erroneous detection.

In addition, by studying the brightness data, scientists can determine a planet's size, the period of its orbit, and even temperatures to determine whether it might be habitable.

While most spacecraft have several instruments, Kepler has just one, called a **photometer**. *Photo* means light and *meter* means measure, so the simplest explanation of Kepler is that it measures the brightness of stars, but with very high precision. The photometer consists of a telescope with one of the largest cameras ever launched into space, a 95-megapixel array of CCDs.

Kepler looked at a 100-square-degree portion of the sky, equivalent in size to two side-by-side "bowls" of the Big Dipper. This star-filled region located between the constellations Cygnus and Lyra contains an estimated 14 million stars, some up to 3,000 light-years away, that are ideal candidates for planet hunting.

Since Kepler is in space and it doesn't have to deal with clouds or the day-night cycle like ground-based telescopes, it can stare continuously at the same star field to constantly monitor any changes in brightness of any of the thousands of stars. Unlike the Hubble Space Telescope, which orbits Earth, Kepler has been placed in what's called an **Earth-trailing orbit** around the Sun, about 60 million miles (96 million km) from our planet.

Kepler provides information about the big picture of exoplanets. While it certainly looks at specific planets, based on the amount and kinds of exoplanets it finds in one region of the sky, scientists can extrapolate the findings for the entire galaxy. This is similar to how surveys or opinion polls work, where a small number of people surveyed represents the opinions of an overall larger population.

The Kepler mission's field of view near the constellation of Cygnus. Credit: NASA

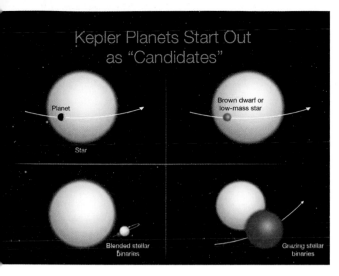

Kepler Planets Start Out
as "Candidates"

Planet

Star

Brown dwarf or
low-mass star

Blended stellar
binaries

Grazing stellar
binaries

Kepler's candidates require verification to determine if they are actual planets and not another object, such as a small star, mimicking a planet. Credits: NASA Ames/W. Stenzel

"Kepler is a statistical mission," Batalha explained. "We are taking a survey of planets in one part of the sky, in one swath, in order to understand the demographics for the whole galaxy of what fraction of stars harbor planets."

As of this writing, Kepler has found 2,325 confirmed planets, with an additional 3,412 planet candidates, meaning they have yet to be confirmed (like the flea mentioned above). Before the Kepler mission, less than 300 exoplanets were known.

In April 2014, Kepler scientists announced the discovery of the first Earth-size planet orbiting a star in the habitable zone. Subsequently, more of these holy grail Earth-size planets were found and confirmed, allowing the Kepler team to estimate the billions of Goldilocks-ready Earth-size planets in our galaxy.

Why search for planets so far away?

Batalha said these planets aren't just a stamp collection, a phrase sometimes used in astronomy to mean science that isn't useful.

"Kepler was designed to answer one very specific question: what is the fraction of stars that harbor potentially habitable Earth-size planets," she said. "The reason we want the answer to that question is so we can continue our pursuit of finding evidence of life beyond Earth. Kepler is one step toward that larger objective."

But the path of getting Kepler into space wasn't easy.

EARLY DAYS OF HUNTING PLANETS

When radio stations started broadcasting in the 1920s, it didn't take long for scientists to realize radio (and later television) waves were broadcasting our transmissions out into space at the speed of light, potentially traveling long distances across the galaxy. They also wondered if the reverse was true: if another civilization from a faraway world was also creating similar transmissions, could we hear them?

In 1959 astronomer Frank Drake came up with the idea for what would eventually become SETI, the search for extraterrestrial intelligence, an active search for alien radio transmissions. So far the search has come up empty, and a small SETI effort continues. However, astronomers have realized with our switch from analog to digital, especially for television, Earth's radio emissions from our technology is waning and might only continue for a short period of time. The same might be true for alien civilizations, too.

The concept of searching for actual planets in the galaxy would prove to be more difficult than just listening for signals since distant worlds are so small, so far away, and don't emit their own light. In the 1960s, astronomers came up with the idea of looking for planets by detecting their effects on the host star. Even the smallest planets induce a gravitational tug on their star, and precise measurements can reveal the presence of a planet in a few different ways.

Astrometry precisely measures the position of stars as compared to other stars around them; any "wobbling" of the star and the extent of that movement can provide information about the planet's mass and orbit.

Radial velocity measures the tiny changes in a star's velocity as the star and the planet move about their common center of mass by measuring the shift in the spectrum of the star's light (called the Doppler shift) as it is either pulled away or toward the observer.

Transits—measuring the changes in brightness of a star when a planet passes in front— provides the most information about a planet, including the radius and period (how long it takes the planet to orbit). If there are multiple worlds in a system, the dynamics of the objects' interactions can provide the planets' masses.

For all methods, the larger the planet and the closer it is to the host star, the greater the change in the star's location, spectrum or brightness. Since the largest and closest planets are the easiest to find, it's not surprising that many of the first planets ever discovered are the size of Jupiter or larger and orbit extremely close to their stars.

This artist's concept depicts the pulsar planet system discovered by Aleksander Wolszczan in 1992 using the Arecibo radio telescope in Puerto Rico to find three planets—the first of any kind ever found outside our Solar System— circling a pulsar called PSR B1257+12. Credit: NASA/JPL-Caltech

The first confirmed discovery of exoplanets came in 1994 using giant telescopes on Earth. Astronomers saw evidence of three planets orbiting a **pulsar**, which is a dense, rapidly spinning remnant of a supernova. Then in 1995 came the first discovery of a planet orbiting a Sun-like star, a giant planet about half the size of Jupiter that goes around its star in a blazing-fast four days. After that, exoplanet discoveries trickled in, and by the year 2000, over 50 exoplanets had been found.

While the discoveries were exciting, in some respects they were a bit of a letdown. All these worlds could not support life as we know it. They either were gas giants or orbited their star so closely that they would be broiling worlds inundated with radiation from the star, leaving them hellish and uninhabitable.

In 2006, the first spacecraft to look for exoplanets was launched by ESA. The Convection, Rotation and planetary Transits (CoRoT) satellite looked for planetary transits like Kepler would later do. In all, CoRoT spotted 32 confirmed planets, with other planet candidates still being verified. The mission ended in 2013 when the spacecraft's main computer failed.

Telescopes on Earth continued their hunt for exoplanets. Some of the most successful ground-based telescope searches include the Keck telescopes in Hawaii, the High Accuracy Radial velocity Planet Searcher (HARPS) at the La Silla Observatory in Chile and HARPS-N on the Canary Islands, Spain.

HISTORY OF THE KEPLER MISSION

The Kepler mission represents a lesson in perseverance. It took five separate proposals until the mission was finally approved by NASA. The crusade to build a space telescope that could look for exoplanets was led primarily by one person, Bill Borucki.

Borucki was a physicist at NASA Ames, and during the 1960s he worked on developing the heat shield for the Apollo missions to the Moon.

He was inspired by NASA's SETI program, which was based at Ames at that time. Borucki was further inspired by a paper published in 1971 by astronomer Frank Rosenblatt, who proposed looking for exoplanets using a new idea, the transit method. The paper wasn't widely read and Rosenblatt died in 1973, so the concept lay dormant. But it was always in the back of Borucki's mind.

Bill Borucki. Credit: NASA Ames

In 1984, he coauthored a paper determining the probability of detecting planets using Rosenblatt's proposed method. Borucki pointed out that ground-based observations should be sufficient to detect Jupiter-size exoplanets, but that the detection of Earth-size planets would require a space-based observatory. He also knew better technology was needed to do the job, so he kept researching the concept and even received some money from NASA to build a proof-of-concept photometer.

Then in 1992, NASA announced its Discovery Program, a new class of lower-cost, highly focused scientific space missions. Borucki and his colleagues submitted proposals in 1992, 1994, 1996 and 1998, but all were rejected due to doubts about the technology and cost.

There was another possible reason for the rejections. Back then the majority of scientists weren't completely convinced that exoplanets were a real thing.

"Basically, exoplanets had not quite reached the status of being accepted by the astrophysics community," said Wesley Traub, NASA's chief scientist for the exoplanet exploration program, "so there were a fair amount of negative thoughts in that direction."

Traub said it was reminiscent of how hard it was to convince astronomers years ago to include the study of the planets in our own solar system in the realm of astrophysics, and to get time allotted on telescopes to focus on planetary science (which might explain why it took until 1978 to discover Pluto's moon Charon).

And some scientists thought looking for other planets—and life—out in the galaxy "borders on the material you see in the supermarket checkout line," Traub said. "So it edges on asking questions that are outside the realm of traditional astronomy. But today, this is the kind of thing that we now are technologically able to do, and there are a lot of people who are really interested in this, both scientists and the public."

Plus there's a big difference between scientifically determining the possibility of life in the universe and thinking you've seen a flying saucer.

But another reason was that NASA didn't (and still doesn't) have a huge budget to work with, and it normally chooses its missions based on the recommendations of what is called a **Decadal Survey,** a report put together by scientists to identify research priorities and make recommendations for missions and funding for the coming decade. An exoplanet mission didn't actually make the list until 2010.

However, Borucki's perseverance convinced all the right people that his concept had merit, and in 2001 NASA funded his Kepler space telescope—named after seventeenth-century astronomer Johannes Kepler, who first described the laws of planetary motion.

"Bill never took those rejections personally, he just kept going because he loves the process of science and discovery," Batalha said about Borucki. "To me, he embodies the essence of NASA— the childlike spirit of discovery, the tireless work ethic and the playful tinkering and risk taking that leads to bold innovation."

BLOWN AWAY

Fast forward to 2009. Kepler launched successfully, and after reaching its orbit, the spacecraft went through its commissioning period, a ten-day checkout of all the systems. In the first observations, scientists saw a tantalizing signal of a potential small planet orbiting a star about 540 light-years

Kepler-10b must be a scorched world, orbiting at a distance that's more than 20 times closer to its star than Mercury is to our own Sun, with a daytime temperature expected to be more than 2,500 degrees Fahrenheit (1,370°C). The Kepler team has determined that Kepler-10b is a rocky planet, with a surface you could stand on, a mass 4.6 times that of Earth, and a diameter 1.4 times that of Earth. Credit: NASA/Kepler Mission/Dana Berry

away. This planet would later be confirmed and named Kepler-10b, the first rocky terrestrial planet found by the Kepler space telescope. This world is about 4.6 times as massive as and 40 percent wider than Earth, but it's not in the star's habitable zone.

"Kepler-10b is one of my personal favorites," said Batalha, "because it was Kepler's first rocky planet. But also because it was one of the first signals that showed not only was Kepler working, but it indicated we were going to find a lot of planets."

But this is an extreme world. Kepler-10b orbits its star closer than Mercury orbits our Sun, and its orbital period, or year, is just twenty hours. Scientists call Kepler-10b a **super-Earth**, a planet that is more massive than Earth but is predominantly rocky.

"What is really amazing about planets like this—and we've since found several of them—one side of the planet always faces the star, and that star-facing side is molten lava," Batalha said. "So there is an ocean of molten rock on one side of the planet, with temperatures in excess of what is required to melt iron."

It was an exciting beginning to the mission. Then, from data collected in just the first 43 days of the mission, the Kepler spacecraft found over 750 planet candidates, an amazing number of potential exoplanets.

"I think everybody was astounded by just how easy it was to find planets in the Kepler data," Barclay said. "Up until that point, each exoplanet discovered was this precious gift that everyone treasured. Then Kepler came along, and we were finding planets every time we looked at the data."

Barclay recalled how the team would have weekly meetings, and one of scientists, Jason Rowe, would plug in his laptop and "show us the goodies that had been given out that week in the data, and we would ooh and aah at the new transits. It was a really exciting time. Everyone in the room was aware that we were seeing something for the first time that was going to change our understanding of our place in the universe."

Also for the first time, solar systems with multiple planets were found—more than a third of planets found were in multi-planet systems. But many didn't look anything like ours. Planets were bunched together in closely spaced orbits, with years that are shorter than a few Earth months.

There were also a few other surprises.

"One of the really interesting things that Kepler found is a new class of planets that turns out to be the most common type of planet known to humanity right now, thanks to Kepler," Batalha said. "These are planets between the size of Earth and Neptune. We call them **super-Earths** or **sub-Neptunes**."

But yet there is nothing like this in our own solar system.

"As an exoplanet scientist, one of the most astounding facts of the entire mission is that we have this size of planets everywhere in Kepler data that are nothing like we have in our own backyard," Barclay said. "In fact before the Kepler mission there was a very famous theory paper that predicted there were no planets about 2.5 times the size of our own Earth. Then we find these are the most common data set from Kepler. And they should be common throughout the galaxy if you take what we observe and turn them into demographics."

One possibility of a world like this in our own solar system, Batalha pointed out, is a currently unseen mystery planet that might be out beyond Pluto. In early 2016, astronomers Mike Brown (the same one who discovered Quaoar, Eris and company) and Caltech colleague Konstantin Batygin proposed a possible large planet (nicknamed Planet Nine) lurking in the far reaches of our solar system. Its presence is inferred because it appears the orbits of several other Kuiper Belt Objects are all being affected by another object that could be up to 10 times more massive than Earth.

This artist's conception of a planetary lineup shows habitable-zone planets with similarities to Earth: from left, Kepler-22b, Kepler-69c, Kepler-452b, Kepler-62f and Kepler-186f. Last in line is Earth itself. Credit: NASA/Ames/JPL-Caltech

An artistic rendering of the distant view from a possible ninth planet in our Solar System, dubbed Planet Nine, looking back towards the sun. The planet is thought to be gaseous, similar to Uranus and Neptune. Credit: Caltech/R. Hurt (IPAC)

Brown said existing telescopes have a shot at spotting this mystery planet because his study points to the region where astronomers should look.

"So, there may be one of these super-Earths in our cosmic neighborhood after all," Batalha said. "It will be interesting to see if this really plays out."

But with this "new" class of planets, Batalha said, so much is unknown. The big question for exoplanet scientists is, what are they all made of?

"Could there be a transitional-type of planet between rocky planets and gas giants," Batalha wondered, "or can you have a planet that is three times the size of Earth that is all a big pile of rock or are they mostly gas? How does it all work? So this is what we are really studying carefully with the Kepler data now."

OF PANCAKES AND PLANETS

Another surprise came in the nature of most planetary systems found by Kepler. They are extremely flat, meaning the orbits of the different planets are all on approximately the same plane, as seen edge on by Kepler.

"If we think of the planetary orbits as kind of a pancake," Batalha explained, "with all the orbits lined up being coplanar, our solar system is like a fluffy pancake with small variations in the alignments of the orbits. But the majority of the systems we're seeing with Kepler are more like a crepe. They are exquisitely flat orbital systems."

Because of the nature of looking for transits, only planets with orbits that by chance align approximately edge on with Kepler's line of sight are visible. With the majority of the systems having very flat, crepe-like orbits, Kepler has been able to see more transiting planets.

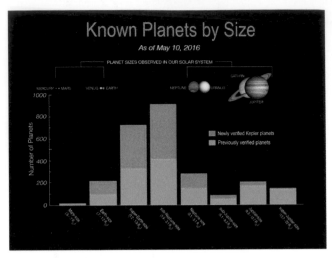

Known Planets by Size
As of May 10, 2016

PLANET SIZES OBSERVED IN OUR SOLAR SYSTEM

MERCURY • • MARS   VENUS •• EARTH   NEPTUNE   URANUS   SATURN   JUPITER

Number of Planets

Newly verified Kepler planets
Previously verified planets

This histogram shows the number of planets by size for all known exoplanets as of May 10, 2016. The blue bars on the histogram represent all previously verified exoplanets by size. Credit: NASA Ames/W. Stenzel

"The volume of discoveries has been really high and that was very surprising," Batalha said. "Really, that's a reflection of the fact that we have discovered this new architecture of coplanar, flat systems."

A BOUNTY OF PLANETS

While the big picture of Kepler's discoveries is certainly impressive, the individual planets and planetary systems are extremely exciting and illuminating.

"We've found so many interesting planets," Batalha said. "My favorites are kind of a 'roll call' of the prototypes that demonstrate the diversity of planets in the galaxy, a diversity that we didn't know existed before Kepler launched."

An artist's concept of the Kepler-16 system, showing the binary star being orbited by Kepler-16b. Credit: NASA/JPL-Caltech

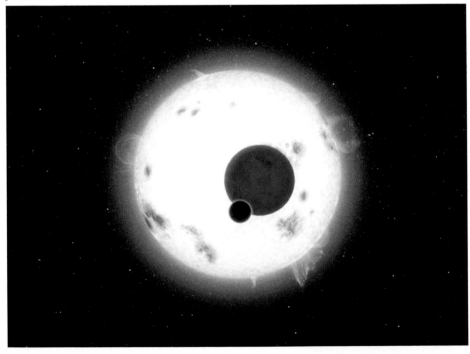

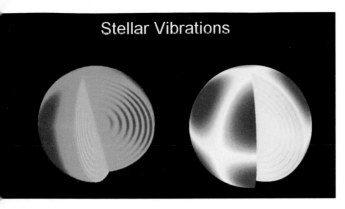

Stellar Vibrations

The variations in brightness can be interpreted as vibrations, or oscillations within the stars, using a technique called asteroseismology. The oscillations reveal information about the internal structure of the stars, in much the same way that seismologists use earthquakes to probe the Earth's interior. Credit: NASA Ames

In 2011, the team announced the discovery of a **circumbinary planet**, a planet that orbits not one but two stars—just like Luke's home planet Tatooine in *Star Wars*.

"If you lived on a circumbinary planet, you'd have two stars rising in the east and setting in the west," Batalha said in wonder. "The stars might change positions as they move across the sky. Exotic places like these have stoked my imagination."

Kepler has also found strange new planets that defy classification, like a world covered in boiling water, a planet getting ripped apart by its star and a Neptune like planet locked in a close orbit along with a rocky world.

In the Kepler data are several examples of planets orbiting extremely old stars, as old as the galaxy itself. Astronomers use a technique called **asteroseismology** with Kepler to basically "listen" to a star by measuring sound waves to determine different parameters of the star, including its age. With this method they discovered an 11-billion-year-old system of five rocky planets, all smaller than Earth. The team says this discovery suggests that Earth-size planets have formed throughout most of the universe's 13.8-billion-year history, increasing the possibility of the existence of ancient life—and potentially advanced intelligent life—in our galaxy.

"This is tremendously interesting," Batalha said, "because it makes me think that there could be civilizations out there that have had more than double the amount of time to evolve. What kind of life might be on those worlds?"

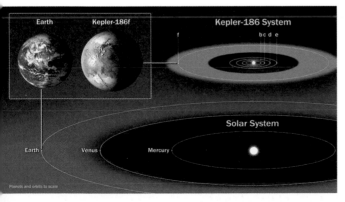

This diagram compares the planets of our inner solar system to Kepler-186, a five-planet star system about 500 light-years from Earth in the constellation Cygnus. The five planets of Kepler-186 orbit an M dwarf, a star that is half the size and mass of the sun. Credits: NASA Ames/SETI Institute/JPL-Caltech

A size comparison of the planets in the Kepler-37 system and objects in our solar system. Credit: NASA/Ames/ JPL-Caltech

Perhaps the most exciting discovery was the first rocky Earth-size world found in a star's Goldilocks zone. While it's too soon to tell if the planet, named Kepler-186f, is really an Earth twin, scientists are now fairly confident Earthlike planets do in fact exist.

"That was a profound moment for the Kepler team," Barclay said, "a watershed moment when we could almost say 'mission accomplished,' as we had changed the face of our understanding forever. This discovery has shown us there are places out there like our own."

Kepler-186f is the fifth and outermost planet discovered orbiting a red dwarf star Kepler-186, located 490 light-years away. The planet completes one orbit around its star every 130 days, just within the outer edge of the system's habitable zone.

Barclay's personal favorite was the discovery of the smallest known exoplanet to date.

"I'm personally proud of this discovery," he said, "because its detection was a challenge. We found a planet smaller than Mercury, so this showed we weren't just discovering these big, hot Jupiters; we were now finding objects that were smaller than the smallest planets in our own system. It really exhibited the diversity of the systems out there."

Kepler-37b is about 210 light-years from Earth, and the planet is slightly larger than our Moon, measuring about one third the size of Earth.

All these discoveries made for exciting times, Barclay said. "We had a large team of people; we had developed a great routine with steady operations, with a tremendous amount of exciting new science coming out. And then suddenly the spacecraft broke."

K2

On May 14, 2013, one of Kepler's reaction wheels failed. These are gyroscope-like devices that keep the spacecraft stable and enable the telescope to be pointed very precisely toward the target field of view. This failure was of grave concern because about a year prior, another reaction wheel quit working. Kepler has four reaction wheels in total, and now two had failed. And the spacecraft needs at least three to be able to point accurately enough to hunt for exoplanets.

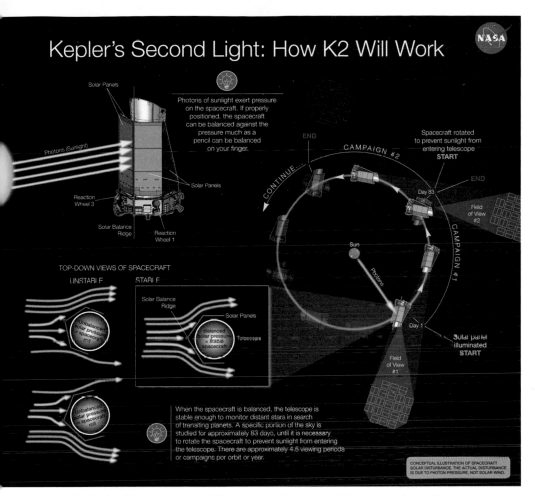

Kepler's Second Light: How K2 Will Work

NASA

Solar Panels

Photons of sunlight exert pressure on the spacecraft. If properly positioned, the spacecraft can be balanced against the pressure much as a pencil can be balanced on your finger.

Photons (Sunlight)

Solar Panels

Reaction Wheel 3

Solar Balance Ridge

Reaction Wheel 1

END

CAMPAIGN #2

CONTINUE...

Spacecraft rotated to prevent sunlight from entering telescope
START

END

Day 83

Field of View #2

CAMPAIGN #1

Sun

Photons

Day 1

Solar panel illuminated
START

Field of View #1

TOP-DOWN VIEWS OF SPACECRAFT

UNSTABLE

STABLE

Solar Balance Ridge

Solar Panels

Unbalanced solar pressure = spacecraft roll

Balanced solar pressure = stable spacecraft

Telescope

Unbalanced solar pressure = spacecraft roll

When the spacecraft is balanced, the telescope is stable enough to monitor distant stars in search of transiting planets. A specific portion of the sky is studied for approximately 83 days, until it is necessary to rotate the spacecraft to prevent sunlight from entering the telescope. There are approximately 4-5 viewing periods or campaigns per orbit or year.

CONCEPTUAL ILLUSTRATION OF SPACECRAFT SOLAR DISTURBANCE. THE ACTUAL DISTURBANCE IS DUE TO PHOTON PRESSURE, NOT SOLAR WIND.

Engineers developed an innovative way to stabilize and control the spacecraft. This technique of using the sun as the "third wheel" has Kepler searching for planets again, but also making new discoveries about young stars and supernovae. Credits. NASA Ames/W Stenzel

"We knew for about six months this was probably going to happen," said Barclay. "It was clear that the second reaction wheel's days were numbered. It was a blow, and we were all pretty down. But many of us were thinking there must be something we can do because we still had this great telescope in space."

The one consolation was even if the Kepler spacecraft was unable to make any more observations, there still were terabytes of data from the mission yet to pore over, where more exoplanets would surely be found. But there was also the chance that the spacecraft could be used for something else that didn't need such quite precise pointing. NASA decided to put a call out for white papers for ideas.

In the meantime, however, an engineer named Doug Wiemer at Ball Aerospace, the company that had built Kepler, was working with other spacecraft with reaction wheels that had failed. This problem has actually affected several missions, especially those on extended missions working longer than their originally proposed prime operations.

Wiemer had come up with a theory involving something called **solar radiation pressure**. When sunlight hits a spacecraft—and in particular when it falls on solar panels that have a large surface area—the photons from the sunlight exert a small but significant force called **radiation pressure**. All spacecraft navigators have to account for this force when they compute their trajectories because this force truly affects a spacecraft's path. It's the force behind the concept of using solar sails to provide propulsion for small spacecraft.

Wiemer's idea involved orienting Kepler so the radiation pressure was evenly distributed on the solar panels (see diagram on page 115) and would act as a third reaction wheel. So, one of the forces the spacecraft normally always fought against would now be used to help stabilize it.

The engineering team at Ball Aerospace tested the idea and it worked.

"It was a brilliant solution," Traub said. "I have to admit, in my role as the chief scientist for the exoplanet exploration program, I spent days and days talking with some of the smartest engineers in the world, but nobody else came up with that idea. It was like what happens in movies or a novel where a miracle appears out of nowhere and saves the day."

While the telescope's aim isn't as precise as it used to be, it's close. The trade-off was that Kepler couldn't point at the same field of view any more. Instead, Kepler's orbit around the Sun would be broken into individual campaigns focusing on a single patch of sky for 83 days, and then after it traveled far enough around the Sun, the radiation pressure would change so the spacecraft would need to be reoriented to compensate. Then it would look at another patch of sky for the following 83 days.

"It was a fantastic solution, an elegant idea for balancing the spacecraft against the solar pressure to keep it stable," Barclay said, "and it worked even better than we expected."

Various targets along the ecliptic during an 83-day K2 viewing time. Credit: NASA Ames

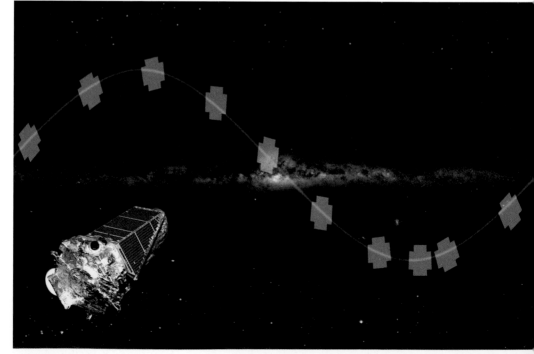

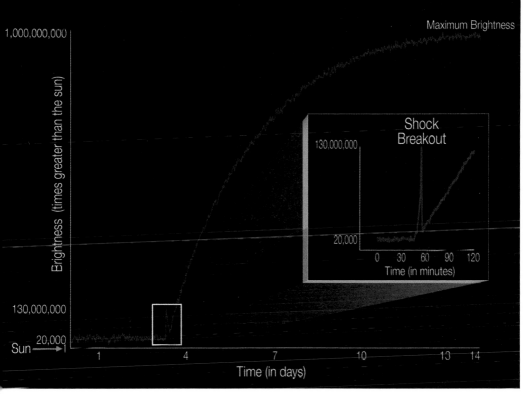

This diagram illustrates the brightness of a supernova event relative to the Sun as the event unfolds. Credits: NASA Ames/W. Stenzel

But since Kepler couldn't look at its original target region in the sky, a new mission emerged called K2.

While the Kepler mission entailed a very focused science goal, K2 is "a completely different beast," Barclay said. "We're not just Kepler slightly worse at pointing; we're a revolutionary new mission."

K2 is now run by a community of scientists around the world, one of the few NASA missions ever not to require a particular science goal. As a community-led mission, a panel of scientists decides how the spacecraft will be used, choosing from submitted proposals. Proposed missions are based on the 83-day region of sky where Kepler will be pointing.

"We are doing some exciting, groundbreaking new science," said Barclay, who now leads the K2 guest observer office at Ames, "science that Kepler actually couldn't do. We're looking for everything from giant black holes to supernovas in distant galaxies. Each time we have a round of proposals, our range gets more and more broad."

Supernovas are the giant explosions of massive stars, but no one knows when they will happen. K2's extended views at a region in space allows for a continuous 83-day lookout for supernovas.

"A supernova takes place approximately once per galaxy per century," Barclay said, "so if you look at 10,000 galaxies at once, you should see a fair number of supernovas go off."

And in March of 2016, the Kepler team announced that for the first time ever, astronomers had captured the brilliant flash of a supernova's shockwave—also called the **shock breakout**—just as the surface of the star erupted in the explosion. The shock breakout itself lasts only about twenty minutes, so astronomers are now able to study what happens just as a star goes supernova. This has never been done until now, and subsequent observations will provide unique insights into these spectacular events. This was actually captured during the prime Kepler mission, Barclay said. "While Kepler cracked the door open on observing the development of these spectacular events, K2 will push it wide open observing dozens more supernovas, and these results are a tantalizing preamble to what's to come from K2!"

Other proposals vie to study the planets in our own solar system since they now fall into K2's field of view. However, exoplanets remain a big part of the mission.

"We're still really an exoplanet powerhouse," Barclay said. "Many scientists moved from Kepler to K2, so we had this built-in community of experts in exoplanet science that were on hand to examine the data rapidly. We're focusing on stars that are much brighter and nearer, stars that are easier to understand and observe from Earth. The idea is to find the best and most interesting systems."

The hope is the K2 mission can continue through 2017. The one limiting factor is how much fuel is left for the spacecraft's thrusters to keep it in the right position.

"It is somewhat difficult to estimate how much fuel we have left," Barclay explained, "and there are two reasons why. One, we only measure tank pressure (not pressure coming out of the thrusters), and two, we aren't sure whether we will be able to drain the tank or whether some fuel will remain. So it is a bit of an art to working out how much longer we can go."

But Kepler might just keep surprising us since it already has an incredible history of overcoming the odds.

"It's been just an amazing, fantastic comeback story of a small team of people who weren't ready to give in and accept the mission was over," Barclay said. "Instead, they created this fantastic science machine to really engage the whole community to do some incredible new science."

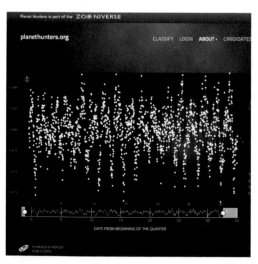

A screenshot from Planet Hunters. As Planet Hunters participants sort through the brightness data in the form of graphs of brightness vs time (known as light curves), they notice different patterns of variability. Much of the variability (on timescales of hours to days) may be caused by starspots or pulsations of different types of variable stars. Having Planet Hunters sort families of similar light curves is part of the important scientific research. Credit: NASA Ames/Zooniverse

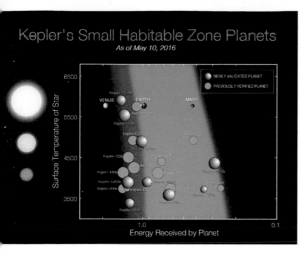

Kepler's Small Habitable Zone Planets
As of May 10, 2016

Since Kepler launched in 2009, 21 planets less than twice the size of Earth have been discovered in the habitable zones of their stars. The orange spheres represent nine newly validated planets announced on May 10, 2016. The blue disks represent the 12 previous known planets. These planets are plotted relative to the temperature of their star and with respect to the amount of energy received from their star in their orbit in Earth units. The sizes of the exoplanets indicate the sizes relative to one another. The images of Earth, Venus and Mars are placed on this diagram for reference. The light and dark green shaded regions indicate the conservative and optimistic habitable zone. Credits: NASA Ames/N. Batalha and W. Stenzel

CITIZEN PLANET HUNTERS

Astronomers aren't the only ones who can search for other planets. You can, too.

A citizen science project called Planet Hunters allows anyone to help Kepler scientists comb through the data and look for signals of exoplanets.

"I think this is tremendously valuable," Batalha said, "for both the Kepler team and for those who participate. It gives people the opportunity to appreciate the scientific process as well as actually make contributions with their discoveries. There is a lot of potential with Planet Hunters—as well as other citizen science efforts—for making valuable discoveries."

Because of the huge amount of data being made available by Kepler, astronomers rely on computers to help them sort through the data and search for possible planet candidates. But the human brain actually works better than a computer in recognizing strange or unique patterns that might indicate a transiting planet. To participate, you don't need any astronomical or exoplanet expertise, and the Planet Hunters' website (www.planethunters.org) is easy to navigate and has interactive tutorials.

So far, Planet Hunters has worked very well, with participants making several discoveries, including finding the first planet ever orbiting *four* Suns. Two citizen scientists from Planet Hunters, Kian Jek and Daryll LaCourse received a special award from the American Astronomical Society for their contributions to exoplanet science.

"I can't stress enough the importance in allowing lay people, and especially young people, to experience the thrill of scientific discovery," Batalha said.

THE FUTURE OF PLANET HUNTING

Bill Borucki retired in 2015, passing the torch on to the next generation to continue the search for other worlds.

"My greatest honor has been the opportunity to develop and lead the Kepler mission. It showed the galaxy is full of Earth-size planets orbiting in the habitable zone of other stars. New and more powerful missions will tell us if the galaxy teems with life," said Borucki. "I hope that young people the world over will take up the challenge to explore our galaxy and will build missions to continue our search for life and to find our place among the stars."

While Kepler has found potentially habitable worlds, how do we figure out conclusively if a distant world is actually habitable and, even more important, if it is indeed inhabited?

Batalha, Barclay and Traub all said Kepler has just scratched the surface of the types of worlds in our galaxy, and the next step is to develop the tools to answer the big question if other life is out there.

"This is an exciting time," Barclay said. "We're right on the cusp of making really big discoveries."

A series of future exoplanet missions are planned, beginning with the Transit Exoplanet Survey Satellite (TESS) scheduled to launch in 2017. The method that TESS will use is identical to that used by Kepler, looking for planetary transits. However, TESS will search stars that are much closer to Earth than Kepler's targets, most of which are 500–3,000 light-years away. Like Kepler, TESS will focus on the identification of Earthlike, rocky planets with the right conditions for liquid water and other constituents hospitable to life. TESS will scan the entire sky to monitor more than half a million stars in our cosmic vicinity.

The best bet for determining the habitability of other worlds might be studying their atmospheres. The highly anticipated James Webb Space Telescope with its 21.3-foot (6.5-m) mirror is expected to do follow-up examinations of the atmospheres of nearby planets found by K2 and TESS, measuring molecules like carbon dioxide, methane and water vapor.

JWST is slated to launch in 2018. It is an infrared telescope (looking beyond visible light), and its primary goals are to study galaxy, star and planet formation in the universe, and to look back in time to see the very first stars and galaxies that formed. JWST is expected to be the premier observatory of the next decade, and you can read more about this mission in the last chapter of this book.

ESA will launch PLATO (PLAnetary Transits and Oscillations) around 2024 to study rocky terrestrial planets in the habitable zone around nearby Sun-like stars, expanding on the use of asteroseismology.

The Wide Field Infrared Survey Telescope (WFIRST) is scheduled to launch in the mid-2020s. The mission will utilize a retrofitted unused spy telescope from the U.S. National Reconnaissance Office that has the same mirror size as the Hubble Space Telescope—8 feet (2.4 m)—but a field of view 200 times wider. This will allow it to cover more sky at greater depth than any previous space observatory. To look for exoplanets, it will use a technique called **microlensing**.

Graphic features NASA's astrophysics missions searching for signs of life beyond Earth. Credits: NASA Ames/N. Batalha and W. Stenzel

"When a foreground star passes in front of a background star," said Traub explaining microlensing, "gravity bends the light and causes it to magnify in the area and makes that background star look brighter. If there is a planet going around the foreground star, it will cause a blip in the light, which WFIRST can measure."

While Kepler was able to see transiting planets in orbits located approximately 1 **astronomical unit** (AU), the distance between Earth and the Sun, and closer to the star, WFIRST will have the sensitivity to detect smaller-than-Earth-size planets orbiting at distances beyond 1 AU. Using a starlight-blocking coronagraph, WFIRST will also be able to directly see reflected light from some larger planets.

"This makes it complementary to the radial velocity and transit detection methods, both of which are most effective at detecting planets orbiting very close to their stars," Traub said. "It's impossible to extrapolate the Kepler data to determine the frequency of planets farther from the star. You have to actually measure things to see what nature produced. You can't just guess because you almost surely will be wrong!"

"Kepler is getting the statistics of exoplanets within an Earth orbit and inward," Batalha said. "WFIRST is going to get the statistics of planets orbiting at an Earth orbit or outward. So, over time, we're going to build up this comprehensive picture of what exoplanets are out there."

Beyond the mid-2020s, the hope is for technology to advance in order to conduct a type of mission that has the power to detect the telltale signs of life in the atmospheres of rocky, Earth-size planets orbiting other stars throughout our stellar neighborhood.

EXISTENTIAL EXOPLANETS

At the beginning of this chapter, Batalha shared the astounding number of perhaps a billion Earth-size planets in the habitable zone of Sun-like stars. But that number is just in our Milky Way galaxy. If you extrapolate that number to the rest of the universe, it's mind-blowing. According to astronomers, there are probably more than 170 billion galaxies in the observable universe stretching out into a region of space 13.8 billion light-years away from us in all directions.

And so, if you multiply the number of stars in our galaxy by the number of galaxies in the universe, you get approximately 10^{24} stars. That's a 1 followed by 24 zeros, or a septillion stars.

This artist's concept depicts planetary discoveries made by NASA's Kepler space telescope. Credit: NASA/W. Stenzel

If other galaxies are like our own, there could be an amazing number of habitable worlds out there. So, while the Kepler mission is important scientifically, it also gives us a glimpse at understanding our humble place in the cosmic ocean.

Batalha once wrote, "As a scientist you live life as if every mystery is there for us to discover and understand." I asked her what it means personally, and perhaps emotionally, to be making discoveries that by their nature help answer questions humans have been asking for centuries.

"There is a certain kind of existential component to doing this work," she said reflectively, "and while it's just a part of my career, I'm afforded the time to think about these broader questions, and I feel tremendously lucky—privileged—to have that in my life. It changes how I approach life and my own thoughts about the human condition.

"Then there is the joy of discovery, the tremendous satisfaction and happiness it brings," she continued. "There is something profoundly joyful and exciting about putting all these puzzle pieces together and making discoveries that change how we look at the world."

Thoughts like these are particularly poignant for Batalha at this time in her life, as she and her family deal with a challenging medical issue for one of her children.

"So, yes, my brain is kind of in a weird place right now," she admitted, "and you really step back and think about the meaning of life. But working on a project like this gives meaning in general because you are doing something that is outside of yourself, outside of our personal problems and struggles, and you really think about the human condition. Kepler really makes us think about the bigger picture of why we're here and what we're evolving toward and what else might be out there."

CHAPTER 6

UNVEILING SECRETS OF A RINGED WORLD AND ITS ICY MOONS: CASSINI-HUYGENS

RUNNING ON EMPTY

It's early 2016 and the Cassini spacecraft is running on fumes.

"The low fuel light is definitely on," said Linda Spilker, the project scientist for the twenty-year-long Cassini-Huygens mission at Saturn. "We just don't know exactly when the bipropellant that helps us shape our orbits might run out."

Despite Spilker's friendly smile and outward calm, the low fuel situation is actually quite worrisome for her and the international Cassini team. The spacecraft needs enough fuel to perform maneuvers for the mission's dramatic but necessary end.

Peaceful beauty: As the Cassini spacecraft approached its target in May of 2004, the spacecraft took this image of Saturn and its rings. A closer look reveals several icy moons as well. Cassini was 17.6 million miles (28.2 million km) from Saturn. Credit: NASA/JPL/Space Science Institute

Artist's concept of Cassini's final orbits between Saturn's innermost rings and the planet's cloud tops. This set of orbits will consist of the last leg of Cassini's mission, called The Grand Finale, culminating with a plunge into Saturn's atmosphere in September 2017. Credit: NASA/JPL

Since 2004, Cassini has been orbiting Saturn, exploring the magnificent gas giant planet while weaving through a startlingly diverse assortment of 62 icy moons and skimming along the edges of the complex but iconic icy rings. Cassini's findings have revolutionized our understanding of the entire Saturn system, which is like its own miniature solar system. Cassini's mission has provided intriguing insights on Saturn itself as well as revealing secrets held by moons such as Enceladus, which should be a big ice ball but instead sports huge hydrothermal geysers. And thanks to the Huygens lander, we now know Saturn's largest moon, Titan, is eerily Earthlike but yet totally alien.

For a mission this big, this long and this unprecedented, it seems fitting to end it in spectacular fashion. In spring of 2017, for her concluding act—called the Grand Finale—Cassini will slip through a small gap between Saturn's cloud tops and the innermost ring, making final but up close observations. Then on September 15, 2017, the spacecraft will plunge down into Saturn and be utterly destroyed by the gas planet's heat and pressure.

In this way, Cassini will conduct a sacred act known as **planetary protection**, ensuring any potentially habitable moons of Saturn won't be contaminated sometime in the future if the drifting, unpowered spacecraft were to accidently crash-land there. Microbes from Earth might still be adhering to Cassini, and its RTG power source still generates warmth. It could melt through the icy crust of one of Saturn's moons, possibly, and reach a subsurface ocean, Spilker said.

Having enough fuel is imperative for the Grand Finale to work. So, for the final months of the mission, whenever the spacecraft performs a trajectory-altering burn, Spilker waits in her office—no matter the hour—until she knows for sure the fuel has held out.

"I just want to watch and wait for that signal that tells me everything worked as planned," she said.

Cassini Project Scientist Linda Spilker and Project Manager Earl Maize receive news that a March 25, 2016 orbital trim maneuver for Cassini was successful, with the spacecraft having enough propellant to complete the burn. Because the exact amount of fuel remaining was uncertain, the spacecraft operations team could not be completely sure Cassini had enough propellant to complete the entire maneuver. Credit: NASA/JPL

Due to Saturn's great distance from Earth, there is a 90-minute one-way delay in communications, and ground controllers cannot give real-time commands or respond to unexpected events.

"We do have contingency plans," Spilker explained, "so if we do run out of fuel partway through a maneuver, we can retool as quickly as we can and use the hydrazine thrusters to complete what we need to do. But those little thrusters would take hours to do what the main engine can do in minutes."

Except for not knowing exactly how much fuel is left, Spilker and the Cassini flight team have everything choreographed to the final plunge.

"We know exactly to the minute what every orbit looks like so we want to stay on the plan and not have to redo it," she said. "If we run out of bipropellant too soon, the science team might have to get together and make some tough trades, and we won't be able to accomplish all the science observations we have planned."

Also, it's a bit unnerving to send the spacecraft you've known and loved for nearly two decades to a potentially dangerous region up close to Saturn, and then intentionally crash it.

"Absolutely, it's scary," Spilker admitted. "We're sending Cassini to a place it hasn't been before, but the scientific return will be incredible. We'll be up so close and making measurements you just can't do anywhere else. And this will help us really—finally—to be able to understand the internal structure of Saturn."

And the team plans to give Cassini a good send-off.

"For that first up close orbit, we'll get the science team together at JPL," Spilker said. "We'll all be holding our breath and waiting for that signal that she's gone through the gap, and hopefully she turns back to Earth and says, 'Yeah I'm OK, no big deal.' And then she'll send back the data, so we can see the exquisite pictures of the rings and the planet itself."

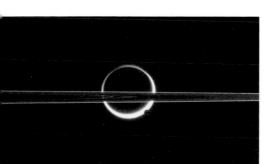

Saturn's rings cut across an eerie scene that is ruled by Titan's luminous crescent and globe-encircling haze, broken by the small moon Enceladus, whose icy jets are dimly visible at its south pole. The scattered light around planet-sized Titan (3,200 miles [5,150 km] across) makes the moon's solid surface visible in silhouette. Enceladus (310 miles [500 km] across) enjoys far clearer skies than its giant sibling moon. Credit: NASA/JPL/ Space Science Institute

Oh, yes … the pictures! The images returned by Cassini have been nothing short of stunning. Incredible shot of the moons, rings and the planet itself are ethereal, art-museum-quality images that can stop you in your tracks and take your breath away.

HOW TO DESIGN A MISSION

With a mission as successful as Cassini-Huygens, it's hard to believe it almost never made it off the ground. Original plans for the mission were canceled, twice. Early ideas in the late 1980s patterned a mission after the earlier and incredibly successful Voyager Grand Tour, which launched in 1977 with twin spacecraft flying past Jupiter, Saturn, Uranus and Neptune. The Voyager encounters sparked so many subsequent questions that a new mission was conceived to focus on Saturn and other solar system bodies.

"At first it was called CRAF/Cassini," said Cassini project manager Earl Maize, "with CRAF standing for comet rendezvous asteroid flyby. The idea was to use two identical spacecraft going different places. It included complex articulating scan platforms for all the instruments and moving antennas. But with all the budget cuts at that time, it was deemed way too expensive."

CRAF was canceled and Cassini was almost canceled, too. But the European and Italian space agencies were teaming up with NASA on the mission, and they had already started building the Huygens probe to study Titan. After much negotiation and in the spirit of international collaboration—in total nineteen countries were working on the mission—Congress gave NASA the go-ahead for Cassini to be built to match up with Europe's Huygens.

"But Cassini was downsized," Maize said, "to the point where everything was just bolted to the body, which changed how the mission operated. The way the spacecraft works is like being on safari and you have your Hasselblad camera bolted on the hood of your Land Rover. If you want to take a picture of that lion over there, you have to change the orientation of the entire vehicle."

But there's not just one camera to deal with. Cassini carries twelve instruments with twenty-seven associated investigations.

"Sometimes the team with the radar instrument wants to point one way while the imaging team wants to point in another direction," Maize said. "Then the fields and particles guys want to point at still another target, and the dust team wants to point a different way. Oh, and then we also have to turn the spacecraft once a day so the high-gain communications antenna is aimed toward Earth. Everyone is contending for their prime attitude."

Maize said the Cassini scientists developed a process—now duplicated by other missions because it works so well—where the instrument teams "get together and slice and dice every piece of the tour" down to the minute. They design and build detailed optimized sequences months in advance, figuring out when each instrument gets that prime position to make its measurements. Once the plan is approved and tested on the ground, ten weeks' worth of observation sequences gets beamed to Cassini, and the onboard computer manages each and every maneuver.

"It's a very careful process, kind of like making a Swiss watch—it all has to integrate perfectly," Maize said. "But it works. And we've been doing it since 2002."

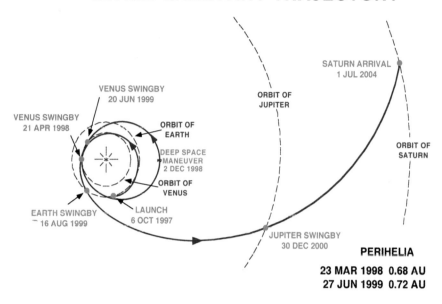

CASSINI - VVEJGA OCT 1997
INTERPLANETARY TRAJECTORY

SATURN ARRIVAL
1 JUL 2004

VENUS SWINGBY
20 JUN 1999

ORBIT OF
JUPITER

VENUS SWINGBY
21 APR 1998

ORBIT OF
EARTH

ORBIT OF
SATURN

DEEP SPACE
MANEUVER
2 DEC 1998

ORBIT OF
VENUS

EARTH SWINGBY
16 AUG 1999

LAUNCH
6 OCT 1997

JUPITER SWINGBY
30 DEC 2000

PERIHELIA

23 MAR 1998 0.68 AU
27 JUN 1999 0.72 AU

This graphic depicts the interplanetary flight path for the Cassini spacecraft, using the VVEJGA (Venus-Venus-Earth-Jupiter Gravity Assist) trajectory. It took 6.7 years for Cassini to reach Saturn. Credit: JPL-Caltech

TROUBLE FOR TITAN

In 1999, the Cassini orbiter and the piggybacking Huygens lander were wending their way to the Saturn system. The duo launched in 1997, but instead of making a beeline for the sixth planet from the Sun, they took a looping path called the Venus Venus Earth Jupiter gravity assist (VVEJGA—gotta love those NASA acronyms) trajectory, swinging around Venus twice and flying past Earth two years later.

While all the flybys gave the spacecraft added boosts to help get it to Saturn, the Earth flyby also provided a chance for the teams to test out various systems and instruments and get immediate feedback.

"The European group wanted to test the Huygens receiver by transmitting the data from Earth," Maize said. "That's a great in-flight test because there's the old adage of flight engineers, 'test as you fly, fly as you test.'"

Artist depiction of Huygens landing on Titan. Credit: ESA

Once at the Saturn system, Huygens would be released from Cassini and drop through Titan's thick and obscuring atmosphere like a skydiver, transmitting data all the while. The Huygens probe didn't have enough power or a large enough dish to transmit all its data directly to Earth, so Cassini would be used as a data relay. Everyone wanted to ensure this data handoff was going to work, otherwise a crucial part of the mission would be lost.

Titan, second only to Earth's Moon in size, has many long-held mysteries. In 1655, astronomer Christiaan Huygens had worked to improve a new-fangled technology called the telescope. With his enhanced lenses, Huygens discovered Saturn had a large, hazy-looking moon. For centuries, astronomers debated whether Titan has an atmosphere like Earth's. No other moon in the solar system has an atmosphere, so could there be life?

When Voyager 1 flew by Titan in 1980, it detected a nitrogen-rich atmosphere kind of like Earth's. But the orange atmosphere was filled with organic smog so thick that Voyager couldn't see anything on the moon's surface. What was down there creating this enshrouding atmosphere? Some scientists predicted oceans, but Titan is so cold—averaging –290°F (–180°C)—there could only be seas of liquid hydrocarbons like methane and ethane. The possibilities of Titan's surface features were almost unimaginable.

The Huygens probe would land on Titan's surface to finally provide a glimpse of this alien world.

"The way it was supposed to work was that Huygens would be flying toward Titan and Cassini would be boring in right behind it, taking the data," said Maize. "So to test it out during the Earth flyby, Huygens, Cassini and the Goldstone Tracking Station of the Deep Space Network were all programmed to simulate the probe descending to Titan. It all worked great."

Except for one thing: Cassini received almost no simulated data, and what it did receive was garbled. No one could figure out why.

Six months of painstaking investigation finally identified the problem. The variation in speed between the two spacecraft hadn't been properly compensated for, causing a communication problem. It was as if the spacecraft were each communicating on a different frequency.

"The European team came to us and said we didn't have a mission," Maize said. "But we put together tiger teams to try and figure it out."

The short answer was that the idiosyncrasies in the communications system were hardwired in. With the spacecraft now millions of miles away, nothing could be fixed. But engineers came up with an ingenious solution using a basic principal known as the Doppler effect.

The metaphor Maize likes to use is this: If you are sitting on the shore and a speed boat goes by close to the coast, it zooms past you quickly. But that same boat going the same speed out on the horizon looks like it is barely moving.

"Since we couldn't change Huygens' signal, the only thing we could change was the way Cassini flew," Maize said. "If we could move Cassini farther away and make it appear as if Huygens was moving slower, it would receive lander's radio waves at a lower frequency, solving the problem."

Maize said it took two years of "fancy coding modifications and some pretty amazing trajectory computations."

However, with Cassini needing to be farther away, it would eventually fly out of range to capture all of Huygens's data. Astronomers instigated a plan where radio telescopes around the world would listen for Huygens's faint signals and capture anything Cassini missed.

Huygens was released from the Cassini spacecraft on Christmas Day 2004 and arrived at Titan on January 14, 2005. The probe began transmitting data to Cassini four minutes into its descent through Titan's murky atmosphere, snapping photos and taking data all the while. Then it touched down, the first time a probe had landed on an extraterrestrial world in the outer solar system.

Because of the communication problem, Huygens was not able to gather as much information as originally planned, as it could only transmit on one channel instead of two. But amazingly, Cassini captured absolutely all the data sent by Huygens until it flew out of range.

The Huygens probe captured this color image of Titan as the spacecraft descended through the moon's atmosphere in January 2005. Credit: ESA/NASA/JPL/University of Arizona

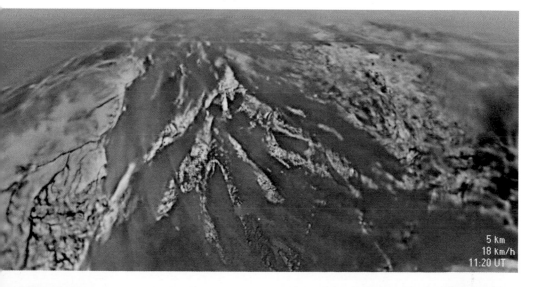

Cassini Mission Ace: Michael Staub, a Cassini mission operations engineer works at the Cassini "Ace" console at the Space Flight Operations Facility (SFOF) at JPL, which he says is like the "cockpit" of the spacecraft. SFOF has monitored and controlled all interplanetary and deep space exploration for NASA and other international space agencies since 1964, and is considered a National Historic Landmark. The control room, also known as the "Dark Room," is filled with huge display screens showing visualizations of data going to and from the distant spacecraft. There are also numerous individual computer consoles for specific space missions. Credits: NASA/JPL, and SFOF image credit: NASA/ Bill Ingalls

"It was beautiful," Maize said, "I'll never forget it. We got it all, and it was a wonderful example of international cooperation. The fact that nineteen countries could get everything coordinated and launched in the first place was pretty amazing, but there's nothing that compares to the worldwide effort we put into recovering the Huygens mission. From an engineering standpoint, that might trump everything else we've done on this mission."

THE MISSION IN SCIENCE AND PICTURES

It took nearly seven years for Cassini-Huygens to get to Saturn, and once there the mission was originally slated to last just four years. Instead, at mission end in 2017, Cassini will have spent over thirteen years at the ringed world, circling it nearly 300 times and witnessing countless fascinating wonders while transmitting extraordinary images and precise measurements back to Earth.

"The lasting story of Cassini will likely be its longevity and the monumental amount of scientific discovery," Maize said. "It was absolutely the right spacecraft in the right place at the right time to capture a huge array of phenomena at Saturn."

Both Maize and Spilker said the longevity of the mission is tribute to those who developed and built the spacecraft as well as the dedicated team who watches over Cassini. The robust spacecraft has encountered very few anomalies during its long tour, and Maize said that while Cassini is very good at diagnosing its own issues, the talented engineering team excels at what they do.

The continuous barrage of discoveries by Cassini has also contributed to the mission's longevity.

"There's something new and exciting every time you turn around, which brings constant reinvigoration to the team," Spilker said.

About 260 scientists from 17 countries have worked on the mission, and Spilker and Maize represent a part of the team that has remained constant for the duration of the mission. But throughout the years there has been an influx of young scientists and engineers who "take their tour of duty with us and then go on to conquer the world somewhere else," Maize said. "Training a whole new cadre of spacecraft engineers and scientists is something I'm pretty proud of, and that helps to keep people revitalized and energized, too."

Cassini's timeline has been jam-packed, with flyby encounters of moons every week and maneuvers in between. "We are busy all the time," Spilker said.

Few sights in the Solar System are more strikingly beautiful than Saturn and its rings. When this image was taken in 2005, the Sun's angle from Cassini's viewpoint created shadows from the rings on the planet's northern hemisphere. Credit: NASA/JPL/Space Science Institute

Bulls-eye moons: Like a cosmic bull's-eye, Enceladus and Tethys line up almost perfectly for Cassini's cameras, and hover over Saturn's rings. Enceladus is 310 miles (500 km) wide and Tethys is 660 miles (1,062 km) wide. Credit: NASA/JPL-Caltech/ Space Science Institute

The long mission means seasons have changed on Saturn during the spacecraft's time in orbit. Cassini has witnessed not quite half a year—nearly two full seasons—at the planet. This has allowed Cassini to view variations in objects—particularly the rings—as the changing sunlight alters viewing angles and temperatures.

POSTCARDS FROM SATURN

The amazing imagery from Cassini has allowed us to travel along with the spacecraft as if we were there to see the incredible views of the rings, moons, geysers and more. While Saturn might be one of the most photogenic spots in the solar system, the cameras and the imaging team deserve credit, too.

The suite of cameras is called the Imaging Science Subsystem (ISS) and consists of a narrow-angle camera that provides high-resolution images of specific targets, and a wide-angle camera that allows a larger field of view at lower resolution.

The leader of the ISS team, Carolyn Porco, has said that what the cameras do is "miraculous, as they convert the fleeting and indifferent fluctuations of electric and magnetic fields into powerful emotion."

With moons and rings sometimes appearing perfectly posed, are all these pictures just lucky shots or well planned-out observations?

"A few of them were just serendipitous, planned for other scientific goals," said Robert West, Cassini imaging team deputy at JPL. But for the most part, just like the science observations, the images are planned out well in advance.

"Those images with multiple moons with Saturn and the rings are mostly created by Mike Evans at Cornell University, working with ISS team member Carl Murray in London," West explained. "Some of them were planned by Carolyn Porco with the breathtaking visual impact in mind."

How does the team know these shots will be available ahead of time?

"We send up thousands of commands every couple of months, and planning where to point the camera begins roughly six months or more before the picture is taken," West said. "For this to work we need to know very precisely where the spacecraft is going to be and where the moons are going to be. This is all 'celestial mechanics'—measuring very accurately from the ground and from previous spacecraft and from Cassini, and calculating using Newton's laws as modified for relativity by Einstein's equations."

With precise calculations—and great software—the spacecraft navigation team is not only able to control Cassini, but also know where the spacecraft and moons are going to be within about six miles (10 km), sometimes years in advance. That knowledge not only creates beautiful images, but accomplishes great science, too. West said the precise determination of moon orbits also provides information about other processes, such as the internal heating that is going on at Enceladus.

Plus, the longer Cassini has been at Saturn, the better the data.

"After accumulating hundreds of such images, it is possible to put all of this information into a computer calculation and solve very accurately for the moon orbits," he said.

Which then leads to even more wonderful images. Here is a look at just a few discoveries and more marvelous images from the Cassini mission.

HUYGENS ON THE SURFACE OF TITAN

"As one of our scientists said, you can only be first once," Maize quipped, "and it was wonderful to be a part of that first landing on Titan."

After the remarkable recovery, Huygens' historic landing on Titan revealed the moon to be curiously like Earth but with an entirely different chemistry. As Huygens descended, it snapped

clear images of Titan's surface from 25 miles (40 km) in altitude, showing spectacular views of bright highlands surrounded by dark plains and canyons. Additionally, there was strong evidence for erosion due to liquid flows and meteorological events like rain. But Titan is so cold—roughly –290°F (–180°C) that the rain isn't water but liquid methane. Measurements of the atmosphere confirmed the presence of a complex soup of organic chemistry, with large amounts of methane and other aerosols. This reinforces the idea that Titan might resemble an early Earth.

Surface of Titan: This colored view shows the landing site for the Huygens probe on Titan. The two rock-like objects just below the middle of the image are about 6 inches (15 cm) across (left) and 1.5 inches (4 cm) across (center), at a distance of about 33 inches (85 cm) from Huygens. The Surface likely consists of a mixture of hydrogen ices, and there appears to be evidence of flurial activity. Credit: NASA/JPL/ESA/University of Arizona.

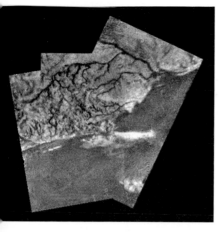

This mosaic of three frames from the Huygens Descent Imager/Spectral Radiometer (DISR) instrument provides unprecedented detail of Titan's surface, showing a high ridge area with a major river channel being fed from different sources. Credit: ESA/NASA/JPL/University of Arizona

A special microphone-like device recorded a "sound spectrum" (though it is not the same as recorded sound) and data revealed "sounds" such as wind, thunder, the deployment of Huygens's braking parachutes and perhaps the splash of methane rain.

Images from the landing site show rounded pebbles that initially looked like a streambed. Scientists now think Huygens landed in something similar to a flood plain on Earth but that it was dry at the time. On first contact with Titan's surface, Huygens bounced and slid, digging a five-inch (12-cm)-deep trench, landing not with a thud but a splat. Further analysis compared Titan's surface to icy, dirty snow that has a frozen crust on top, and when Huygens bounced down, it broke through the crust and sank in.

Huygens transmitted data during its 2-hour-and-27-minute descent and sent a signal for an additional 72 minutes on the surface, much longer than anticipated.

"It was amazing that Huygens survived the landing and lasted so long," said Maize. "There weren't any glitches, even at impact."

Huygens provided the first real view of Titan, showing geological and meteorological processes that are more similar to those on the surface of Earth than anywhere else in the solar system.

ACTIVE, ICY GEYSERS ON SATURN'S MOON ENCELADUS

"One of the biggest surprises of the mission is Enceladus, a tiny moon with active geysers at its south pole," said Spilker. "When Cassini launched, we had no idea we would find an active world circling Saturn."

At only about 310 miles (500 km) in diameter, the bright and ice-covered Enceladus should be too small and too far from the Sun to be active. Instead, this little moon is one of the most geologically dynamic objects in the solar system.

Stunning backlit images of the moon show plumes erupting in Yellowstone-like geysers, emanating from tiger-stripe-shaped fractures in the moon's surface. The discovery of the geysers took on more importance when Cassini later determined the plumes contained water ice and organics. Since life as we know it relies on water, this small but energetic moon has been added to the short list of possible places for life in our solar system.

While pictures of Enceladus' plumes are incredibly striking, it was another instrument on Cassini, the magnetometer, that first noticed something strange on the first flyby of Enceladus.

"Usually with an airless body," explained Spilker, "Saturn's magnetic field lines reach the surface. However, during the first two close Enceladus flybys, the magnetometer team reported a 'draping' of Saturn's magnetic field around Enceladus instead, suggesting an atmosphere. This new discovery convinced us to fly much closer with the next flyby."

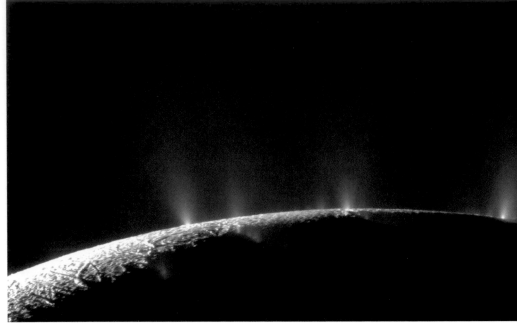

Dramatic plumes, both large and small, spray water ice and vapor from many locations along the famed "tiger stripes" near the south pole of Saturn's moon Enceladus. The tiger stripes are four prominent, approximately 95-mile (150-km)-long fractures that cross the moon's south polar terrain. Credit: NASA/JPL/Space Science Institute

Subsequent observations with Cassini's Composite Infrared Spectrometer showed Enceladus's south pole was much warmer than expected, with heat radiating from the entire length of several 95-mile (150-km)-long fractures, at more than twice the power output of all the hot springs in the Yellowstone region on Earth.

The Cassini team identified over 100 active geysers—sometimes called **cryovolcanoes**—meaning there had to be a huge reservoir of liquid water in Enceladus' interior. Scientists now predict a global six-mile (10-km)-deep ocean of water larger than Lake Superior in the Arizona-size moon.

"A global ocean indicates it has been there a long time, probably since Enceladus was formed, and recently Cassini discovered hints of hydrothermal activity," Spilker said. "We now have all this evidence of an environment that might be conducive to life, so we'd certainly like to go back someday with a dedicated mission to study Enceladus and answer all these life-related questions."

Since the geysers weren't known until Cassini's arrival at Saturn, the spacecraft didn't have a specific payload to study them. But the team made the most of the very capable and diverse set of instruments, taking the spacecraft on several close encounters—coming within 30 miles (49 km) of the surface—and passing *through* the icy plumes. This allowed the instruments to "taste and smell" the particles in the plumes.

"This discovery really reshaped what our focus was for the entire tour," said Maize. "We went from having about 3 flybys to doing a total of 22. We were able to observe this moon again and again, and that's the legacy of a prolonged mission like Cassini."

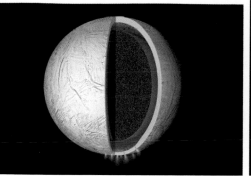

Illustration of the interior of Saturn's moon Enceladus showing a global liquid water ocean between its rocky core and icy crust. Credit: NASA/JPL-Caltech

Saturn casts a big shadow on its rings in this natural color mosaic from Cassini taken in 2007. Three moons are also visible in this image: Mimas (247 miles [397 km] across) at the 2 o'clock position, Janus (113 miles [181 km] across) at the 4 o'clock position and Pandora (52 miles [84 km] across) at the 8 o'clock position. Pandora is a faint speck just outside the narrow F ring. Credit: NASA/JPL/Space Science Institute

What generates the internal heat at Enceladus? The main source of heat remains a mystery, but scientists think gravitational forces between Enceladus, Saturn and another moon, Dione, pull and flex Enceladus's interior. Known as **tidal forces**, the tugging causes the moon's interior to rub, creating friction and heat.

The output of ice and particles from Enceladus is voluminous enough that it created an additional ring around Saturn. "Enceladus has created the E ring, the widest and outermost ring of the main Saturn system," Spilker said. "This is very exciting, especially to a scientist like me who has a tremendous interest in planetary ring formation. There aren't a lot of active objects in the solar system, and to have one create a planetary ring is quite compelling."

SATURN'S SURPRISINGLY ACTIVE AND DYNAMIC RINGS

The striking and mysterious rings of Saturn are among the most recognizable features in the solar system. Even when viewed through a small telescope from Earth, the sight of the rings will take your breath away. Far from being a massive single object, the rings are made of millions of particles in various shapes and sizes. Cassini's long mission has made it possible to watch up close the changes in Saturn's dynamic ring system, and they've now become a laboratory for watching how planets might form.

"Cassini has changed our paradigm for planetary rings," Spilker said. "We've gone from a simple view of thinking the individual particles might just gently bump into each other occasionally, to realizing that most of Saturn's main rings, particularly the A and B rings, have particles that clump together in structures and form gravity wakes. This behavior is giving us a better understanding of how the ring material interacts and also how rocky debris in the early solar system may have come together to form planets."

Spilker said that the structures in the rings are sometimes ephemeral. They come together and then float apart, but they tend to try and reform again.

Saturn's rings start just 4,400 miles (7,000 km) above the surface of the planet and extend out to 175,000 miles (280,000 km) across, meaning Saturn and its rings would fit nicely in the space between Earth and the Moon. The rings are composed of chunks of water ice with a sprinkling of rocky material that range from the size of apartment buildings to grains the size of talcum powder. They circle Saturn at speeds of 20,000 to 24,000 miles per hour (32,190 to 38,620 kph).

But the rings are gossamer thin, measuring less than 30 feet (10 m) thick. "If you put the entire mass of the rings together, it would be no bigger than the moon Enceladus," said Spilker. "So we get this tremendous visual spectacle for comparably little mass!"

Despite its meager mass, the ring system contains an enormous variety of structures. Cassini discovered mini moons that create weird shapes in the ring particles, witnessed the possible birth of a new moon, watched meteor impacts on the rings and observed what may be one of the most active, chaotic rings in our solar system, the F ring.

The rings are named alphabetically in order of their discovery, so the order of the main rings outward from Saturn is a bit confusing: D, C, B, A, F, G and E. "This is an example of astronomers having no imagination whatsoever when it comes to naming things," Spilker said with a smile.

We've known about the rings since 1610, when Galileo turned one of the first telescopes on Saturn, seeing what he described as handles attached to the planet, thinking it was perhaps big moons on either side. Christiaan Huygens later figured out that the handles were actually rings, and in the 1670s, Italian astronomer Giovanni Cassini was able to resolve the rings in more detail, even observing gaps in the rings. One gap is called the Cassini Division and the mission was named after him.

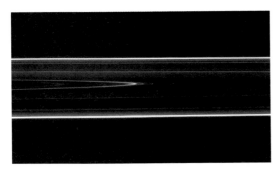

Saturn's rings, labeled with their names. Credit: NASA/JPL/Space Science Institute

As a ring scientist, Spilker has studied the composition of the ring particles. She and fellow ring scientists know they contain mostly water ice, but there are very subtle signatures of other components they still don't understand. "Near the end of the mission we will fly near the inner edge of the rings and directly sample the ring particles. In this way we hope to be able to figure out their composition."

Spilker has envisioned holding a ring particle in her hand. What would it look like?

"We have evidence of the particles that have an icy core covered with fluffy regolith material that is very porous," she said, "and that means the particle can heat up and cool down very quickly compared to a solid ice cube."

It's quite fascinating that as the seasons have changed at Saturn, Cassini viewed the ring particles go through phases just like the Moon.

"As the Sun changes location, you might see one side of the particle that is fully illuminated, and as the Sun changes position, the particles move and you get the 'phases of the Moon' effect if the particles are large enough," Spilker said. "From that, we get information about their spin and are able to measure their temperatures."

Vertical structures cause shadows on Saturn's B ring in this August 2009 picture from the Cassini spacecraft. Credit: NASA/JPL/Space Science Institute

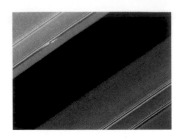

A propeller-shaped structure created by an unseen moon is brightly illuminated on the sunlit side of Saturn's rings. The moon, which is too small to be seen, is at the center of the propeller structure visible in the upper left of the image, near the Encke Gap of the A ring. The A ring is the outermost of Saturn's main rings. Credit: NASA/JPL/Space Science Institute

Changing Sun angles helped reveal towering vertical structures in the planet's otherwise flat rings that form from the gravitational effects of small nearby moons, and scientists have been able to watch how these particles interact with each other and how wakes, eddies and gaps form in the rings.

When Cassini scientists noticed some weird-looking propeller-shaped gaps in the outer edge of Saturn's A ring, a closer look revealed they were being formed by dozens of moving moonlets. The mini moons range between one and several km in diameter, too small to be imaged directly by Cassini's cameras so are only distinguishable by the unique double-armed propeller features they create. This highlights a new understanding: Saturn's rings are always changing.

Where did Saturn's rings come from? That's been a question many scientists have been trying to answer.

"Saturn's rings may have formed from the breakup of a large object such as a moon or comet, or perhaps they formed at the same time as Saturn formed," Spilker said. "However, any theory about the origin of Saturn's rings must also explain the large water ice makeup observed by Cassini."

Spilker said a recent model speculates that a large Titan-size object spiraled into a young Saturn and lost its upper icy layers while its sturdy rocky core plummeted into the planet. These fragments would produce a massive water ice ring. Over time, some of the ice particles would clump near the outer edge of the ring and spin off to form icy moons such as Enceladus, Tethys and Dione.

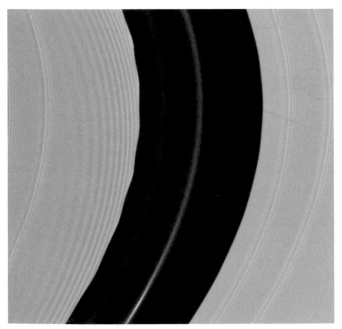

This detailed view of the region in Saturn's rings known as the Encke Gap, a 200-mile (322-km) void in Saturn's outermost main A ring. Cassini scientists determined a small moon named Pan maintains the Encke gap by a "shepherding" mechanism, clearing it of ice and dust and leaving a scalloped edge and a wake pattern on the left that spreads towards the giant planet. Credit: NASA/JPL/Space Science Institute

This series of images from Cassini shows the development of the largest storm seen on the planet since 1990. These true-color and composite near-true-color views chronicle the storm from its start in late 2010 through mid-2011, showing how the distinct head of the storm quickly grew large but eventually became engulfed by the storm's tail. Credit: NASA/JPL/ Space Science Institute

The age of the rings has also been a long-standing mystery.

"The more massive the rings, the longer they have been around," Spilker said. "Less massive, and perhaps a moon, comet or asteroid broke apart and formed the rings only a few hundred million years ago. For understanding the age and evolution of the rings, obtaining a good measurement of the ring mass as a whole is key."

There are estimates now, but Spilker said with Cassini's final orbits, they should be able to get verification when the spacecraft goes inside the rings. "We'll be able to determine the mass of Saturn alone, and then we can subtract the mass of Saturn, and the rings together to obtain an accurate measurement of just the mass of the rings."

STUDYING SATURN'S SUPERSTORMS

Saturn is about 10 times larger than Earth and sports bigger storms, too. Late in 2010, Saturn's relatively calm atmosphere erupted with a storm of gigantic proportions. It lasted for months and grew to encircle the planet with a swirling band. It was a thunder-and-lightning-type storm (no rain, however) that endured for 201 days, creating a huge swirling vortex that expanded to 7,500 miles (12,000 km). The storm caused the largest temperature increases ever recorded for the upper atmosphere of any planet. Scientists say a storm like this develops once every Saturn year, about 30 Earth years, so it was lucky timing that allowed Cassini to watch the storm develop and eventually subside.

Two other storms on Saturn are continuous, monstrous and mysterious tempests on both of Saturn's poles. The south pole has a giant hurricane-like storm with a huge, dark eye ringed by towering clouds. This Saturnian hurricane spans approximately 5,000 miles (8,000 km) across, or two thirds the diameter of Earth, with winds clocked at around 350 miles per hour (550 kmh). This giant storm differs from hurricanes on Earth because it is locked to the pole and does not drift around. Also, since Saturn is a gas planet, the storm apparently forms without an ocean at its base.

At the north pole churns a massive hexagon-shaped storm about 20,000 miles (30,000 km) wide, with equally high winds (340 mph [550 kmh]) as the south pole's. Powered by a band of upper-atmospheric winds, oddly, the interior whips around at high speeds, while the outer hexagon region doesn't seem to move.

This spectacular, vertigo inducing, false-color image from NASA's Cassini mission highlights the storms at Saturn's north pole. Credit: NASA/JPL-Caltech/Space Science Institute

The driving forces of each storm remain a mystery, and being able to watch the storms during seasonal change with Cassini data will help scientists understand the dramatic meteorology taking place at the poles of Saturn.

MYSTERIOUS MOONS REVEALED

The dozens of icy moons orbiting Saturn vary drastically in size, shape and surface features. Some of them have hard, cratered surfaces; some have cracked, icy exteriors; whereas others are, well, just weird.

Hyperion might be the strangest of all, as it looks like a giant sponge. It measures just 168 miles (270 km) across, and its unusual appearance is attributed to the Moon's low density and weak surface gravity, making it highly porous. Cassini flybys have revealed that 40 percent of Hyperion consists of, remarkably, empty space. It is mostly water ice, not rock.

Hyperion also rotates chaotically as it orbits Saturn. It wobbles so much on its axis that Cassini scientists say the Moon's orientation is unpredictable.

Sponge or moon? Cassini scientists think that Hyperion's unusual appearance can be attributed to the fact that it has an unusually low density for such a large object, giving it weak surface gravity and high porosity. Credit: NASA/JPL/Space Science Institute

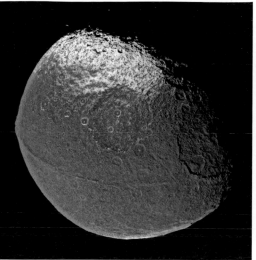

Saturn's moon Iapetus with its remarkable topographic ridge that coincides almost exactly with the geographic equator. Credit: NASA/JPL/Space Science Institute

Mimas, with its large Herschel Crater, makes the moon look like the Death Star in the movie Star Wars. *Credit: NASA/JPL/Space Science Institute*

Iapetus, nicknamed the yin-yang moon for its two different-colored sides, has fascinated astronomers for more than 300 years. Cassini determined the dark reddish dust gathered on one side as the moon plowed through debris in its orbital path. The other, "clean," side remains bright icy white.

Iapetus is nearly 1,000 miles (1,609 km) wide, and it also has a weird ridge of mountains at its equator, making the moon look like a space walnut. Scientists have several ideas about how the ridge formed, but one theory says a collision between Iapetus and another planetary body generated a large quantity of debris. This debris settled into orbit around Iapetus's equator and eventually rained down from orbit, piling up to form the equatorial mountains.

The moon Mimas, nicknamed the Death Star for its resemblance to the *Star Wars* moonlike space station, averages 246 miles (396 km) in diameter, and the giant crater that creates the bull's-eye is 80 miles (130 km) across. Cassini scientists say that if the object that struck Mimas to create this crater had been larger or been moving faster, Mimas would probably have been "disrupted" into pieces to create another ring of Saturn.

Phoebe is another weird moon. It's about 130 miles (220 km) wide, irregularly shaped and quite dark. It orbits Saturn in a retrograde motion, the opposite way of the other moons and Saturn itself. This leads scientists to think Phoebe is likely a wandering object from the Kuiper Belt that was later captured and pulled into orbit by Saturn. Phoebe creates a dusty ring that moves closer to Saturn and provides the dust that coats the dark side of Iapetus.

Other moons, like Dione and Tethys, show evidence of tectonic activity, where forces from within have ripped apart their surfaces. Many, like Rhea and Tethys, appear to have formed billions of years ago, while others, like Janus and Epimetheus, could have originally been part of larger bodies that broke up. Scientists say the study and comparison of these moons tells us a great deal about the history of the Saturn system, and our solar system, too.

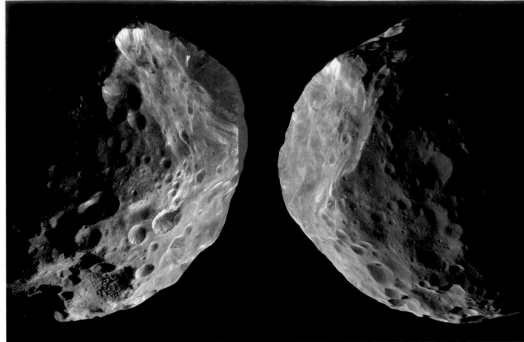

This montage of two views of Saturn's moon Phoebe shows Cassini's view as it approached the moon on the left, while the right side shows the spacecraft's departing perspective. Credit: NASA/JPL-Caltech/Space Science Institute

The existence of oceans or lakes of liquid methane on Saturn's moon Titan was predicted for decades, but the moon's dense haze prevented a closer look. Until the Cassini flyby of July 22, 2006, that is. Radar imaging data from the flyby provided convincing evidence for large bodies of liquid. Credit: NASA/JPL-Caltech/USGS

LAKESHORE PROPERTY ON TITAN

While the Huygens probe revealed much about Titan, continuing studies with Cassini's infrared mapping spectrometer and radar instruments that can peer through Titan's dense atmosphere has revealed an even more Earthlike world.

"Titan has just continued to amaze us for the entire mission," Maize said. "It looks to have a global subsurface liquid water ocean, as well as lakes, rivers, sand dunes, plus an entire weather system with methane rain, but just a completely different chemistry than Earth."

"Titan has many geological processes similar to that of Earth's, and these processes generate methane rains, filling Titan's hydrocarbon lakes and seas," Spilker said. "Titan is the only place in the solar system besides Earth that we currently know has stable liquid on its surface."

In this near-infrared global mosaic of Titan, sunglint and the moon's polar seas are visible above the shadow of nighttime darkness. Credit: NASA/JPL-Caltech/University of Arizona/University of Idaho

Cassini's cameras captured a rare view of Saturn's rings and our planet Earth (bright dot below the rings on the right) and its Moon in the same frame. NASA combined images taken using red, green and blue spectral filters to create this natural color view, as it would appear to the human eye. Cassini scientists asked people on Earth to wave at Saturn at the approximate time the images were being taken. In the photo, Cassini was 753,000 miles (1,211,836 km) from Saturn and 898,414 million miles (1,445,857 million km) away from our planet. Credit: NASA/JPL-Caltech/Space Science Institute

Some lakes on Titan are as big as the Great Lakes in the United States or the Caspian Sea, which lies between Europe and Asia. Cassini has even captured sunlight glinting off one of the lakes in Titan's northern hemisphere, with hints of waves on some lakes.

Cassini's instruments have also seen evidence of dry lake beds at the south pole, and the current theory is the lakes fill and then dry out again, with the rains and liquids transitioning from hemisphere to hemisphere during the 30-year seasonal cycle on Titan.

"One cool thing we've done is bounced radio signals off the lakes—kind of like shining a flashlight on a puddle," said Maize, "but you can actually plumb the depths of the lake with radio signals."

Kraken Mare, the biggest hydrocarbon sea on Titan, is at least 115 feet (35 m) deep and perhaps much deeper.

"Every time we fly by Titan, we see something new," Maize said. "I'm just fascinated by it."

CASSINI'S GRAND FINALE

When I talked with Spilker in her office at JPL in early 2016, Cassini had recently completed its final flyby of Enceladus.

"It's sad to realize we're doing some of 'the lasts' now with Cassini," she said. "There is definitely a sadness as the mission's end draws near, but at the same time a tremendous sense of accomplishment. Cassini has been more successful in many ways than anyone could have imagined when we launched, with the mission lasting so long with such a healthy spacecraft."

"In one way," Spilker continued, "the mission will end. But we have collected this treasure trove of data, so we have decades of additional work to do. With this firehose of data coming back basically every day, we have only been able to skim the cream off the top of the best images and data. But imagine how many new discoveries we haven't made yet. The search for a more complete understanding of the Saturn system continues, and we leave that legacy to those who come after, as we dream of future missions to continue the exploration we began."

CHAPTER 7

DOWNLOADING THE SUN 24/7: THE SOLAR DYNAMICS OBSERVATORY

CLOSE CALL

On July 23, 2012, a rapid succession of powerful explosions erupted on the surface of the Sun. These explosions, called **coronal mass ejections** (CMEs) were equal to the energy of thousands of nuclear bombs detonating at once. The explosions sent blasts of billion-ton clouds of magnetized plasma hurtling out into space at 1,900 miles per second (3,000 km per second), four times faster than a typical solar eruption.

This was one of the strongest solar storms in recorded history.

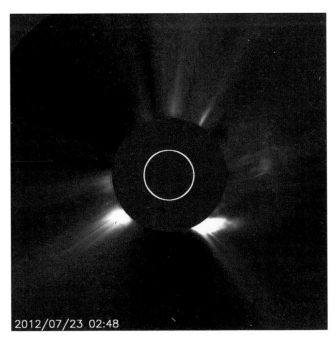

This image was captured by ESA/NASA's Solar and Heliospheric Observatory (SOHO) in July, 2012. On the right side, a cloud of solar material ejects from the Sun in one of the fastest coronal mass ejections (CMEs) ever measured. Credits: ESA&NASA/SOHO

2012/07/23 02:48

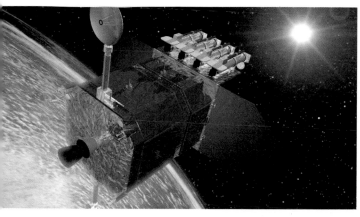

An artist's concept of the Solar Dynamics Observatory in space. Credit: NASA

Would this mega solar storm hit Earth? If so, it could cause global power blackouts, blowing transformers and frying everything plugged into an outlet. Communications around the world could be severed as satellites for phones, televisions, radios, the Internet and military applications might be incapacitated and possibly tumble out of control. Astronauts on the International Space Station would be in danger of being exposed to dangerous levels of radiation from the solar particles, and systems on board could be damaged. The damage from such a solar storm could take years to repair, and in the United States alone, the total economic impact could exceed $2 trillion or 20 times greater than the costs of an event like Hurricane Katrina. Our technology-dependent society would be crippled.

Fortunately, a fleet of solar observatories was watching the action as it happened, and astronomers knew Earth wasn't in the line of fire. This was a close call, however. If the CME had happened just days earlier, our planet could have experienced a global disaster unlike anything we've ever experienced.

Being able to keep a constant eye on the Sun is why scientists built NASA's Solar Dynamics Observatory (SDO). This spacecraft allows us to see the Sun like never before.

"SDO studies the Sun almost constantly and looks at the solar storms that drive our space weather here on Earth," said Tom Woods, principal investigator of one of the instruments on board SDO called the Extreme Ultraviolet Variability Experiment. "Learning more about solar storms is especially important because these events can have an impact on a lot of our technologies, such as communication, GPS, navigation systems. And one of the key goals for SDO is to understand better the causes for these solar storms and how to better forecast them."

SDO is unlike any other Sun-watching spacecraft or ground-based solar observatory because of the quality of the instruments and the staggering amount of data this mission produces. Six high-resolution (4k by 4k pixel) cameras capture images of the Sun every 0.75 seconds, with 10 times better resolution than high-definition television. SDO creates stunning, never-before-seen movies of the solar surface, showing intricate detail. This allows astronomers to see features and events on the Sun that weren't known until SDO was launched into orbit in 2010.

SDO has also been a trailblazer for space missions that produce huge amounts of data. Three hard-working science instruments on board the spacecraft collect a staggering 1.5 terabytes of data *every day*. That would be the equivalent of filling a music CD every 36 seconds or downloading 380 full-length movies or 500,000 songs—each day.

Before we learn more about SDO, let's take a closer look at the Sun.

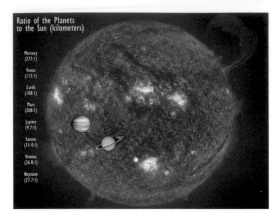

Ratio of the Planets to the Sun (kilometers)

Mercury (277:1)
Venus (113:1)
Earth (108:1)
Mars (208:1)
Jupiter (9.7:1)
Saturn (11.4:1)
Uranus (26.8:1)
Neptune (27.7:1)

How big is the Sun compared to the planets? This infographic compares the ratios of the size of the Sun compared to all the planets in our Solar System. Credit: NASA/JPL

HOW DOES THE SUN WORK?

There's a reason why our neighborhood in space is called the "solar" system. Our Sun is the hub of the solar system, literally and figuratively. It contains 99.9 percent of all the matter in our solar system and is the source of all life and energy here on Earth. The Sun's gravity dominates all the planets and objects that orbit it. Since the beginning of history, human beings have seemingly always understood the Sun's importance to our world and how it shapes the sequence of day and night, the seasons and the cycle of life.

About 4.6 billion years ago, a cloud of gas and dust in space came together after the explosion of a star—called a **supernova**—and the cloud began to collapse, forming a solar nebula. Just like skaters who spin faster as they pull in their arms, the cloud began to spin as it collapsed, forming a hot, dense center that eventually coalesced into a star, our Sun.

The solar system formed from the remaining gas and dust that spun around the newly forming star. Thanks to the Sun's massive gravity, the planets, asteroids, comets and other objects still continue this rotating course.

Stars, including the Sun, don't burn. Instead, the insides of stars could be compared to perpetually exploding hydrogen bombs or mega fusion reactors. Stars collapse under their own weight, crushing the interior and building up enormous pressure. Because of this pressure, hydrogen atoms are fused together to create helium in a process called nuclear fusion. At its center—called the **core**—the Sun fuses millions of tons of hydrogen each second, creating huge amounts of continuous energy. While it can take millions of years for the energy produced in the Sun's core to reach its surface, once in space that energy takes only about eight minutes to travel the 93 million miles (150 million km) to Earth.

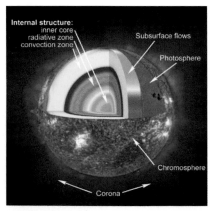

Internal structure:
inner core
radiative zone
convection zone
Subsurface flows
Photosphere
Chromosphere
Corona

Image of the Sun with cut-away portion showing the different layers. Credit: NASA

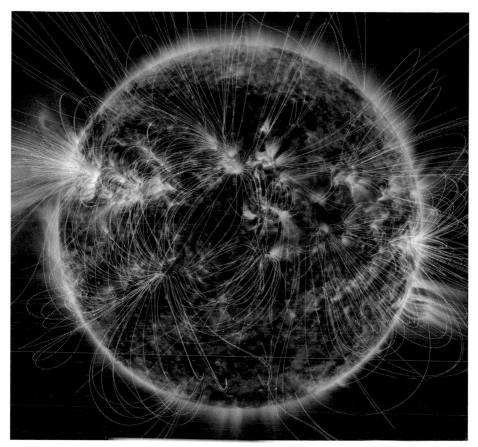

This image shows a model of the magnetic field in the sun's atmosphere based on magnetic measurements of the solar surface. The underlying image was taken in extreme ultraviolet wavelengths of 171 angstroms. This type of light is invisible to our eyes, but is colorized here in gold. Credits: NASA/SDO/AIA/LMSAL

The Sun can be divided into two different regions—the interior and exterior—each with three layers. As mentioned, the core is where all the action begins, with fusion generating temperatures of 25,000,000°F (14,000,000°C). Then comes the radiative zone, where heat slowly rises from the core, taking up to millions of years to move through this region. In the third layer, the convective zone, the heat begins to move quickly—churning and boiling—and it takes just a month for the heat to rise to the surface level of the Sun.

In the exterior region, the first layer is the visible surface known as the photosphere, where astronomers can see features like sunspots. Here, temperatures have "cooled" to about 10,000°F (5,500°C). The next two layers could be considered the Sun's atmosphere: first is the chromosphere, an active region where we see filaments and prominences erupt outward from the Sun, and where temperatures—oddly enough—begin to climb and can quadruple from the photosphere, up to 50,000°F (20,000°C.) The second outermost layer, the corona, is where gases are superheated to temperatures greater than 1,800,000°F (1,000,000° C). The Sun is so hot that most of the gas is actually plasma, a gas-like state of matter in which electrons and ions have separated, creating a superhot mix of charged particles.

One of the biggest mysteries in solar science is why the temperature increases with height in the chromosphere and corona. Scientists suspect the magnetic activity in the Sun's atmosphere provides the energy behind the heating.

Understanding the Sun's magnetic field is key to understanding several aspects of solar activity; however, scientists don't have a full picture of it yet. They do know that when the charged particles in the Sun's plasma move, they naturally create magnetic fields that twist and loop. When these fields interact, they can unleash the energy in the form of CMEs, and **solar flares**, which are smaller events than CMEs but they travel at light speed, so they can reach Earth in less than eight minutes.

The magnetic field on the Sun goes through a regular cycle of activity every eleven years, going from what is called **solar maximum**, where sunspots, flares and CMEs occur frequently, to **solar minimum**, where the Sun is relatively quiet. This regular rhythm has been going on for millennia, but, again, scientists are still trying to understand it fully.

As further evidence for the influence the Sun has on the entire solar system, the Sun's hot corona continuously expands in space, creating what is called the solar wind, a stream of charged particles that extends out 100 times the distance of the Earth from the Sun—out past the farthest reaches of our solar system—to a region called the heliopause. An enormous bubble formed by the solar wind, called the heliosphere, is the largest continuous structure in the solar system.

Astronomers know that stars like our Sun shine for 9 to 10 billion years. With its current age of about 4.6 billion years, our Sun is at its prime and has over 5 billion years of continuous shining yet to do before it begins to die and expand into a red giant star.

HOW SDO WORKS

Most of the time—to our eyes—the Sun appears constant and unchanging. As it moves across the sky every day, we may hardly give it a second thought. While we know the Sun is responsible for and greatly influences our life on Earth, it still holds many mysteries. We don't understand completely how the inside of the Sun works or how energy is stored and released in the Sun's atmosphere. We also don't fully understand why solar activity such as solar flares and CMEs occur.

SDO is helping us understand all this by observing the Sun on both small and large scales, as well as in multiple wavelengths beyond what our eyes can see.

One particular goal of SDO is to understand how the Sun's magnetic field is generated and structured, and how this stored magnetic energy is converted and released into space in the form of solar wind and energetic particles. Getting a handle on what drives the magnetic field is crucial for comprehending how the Sun affects Earth and the entire solar system.

LAUNCH DAY: FEBRUARY 11, 2010

At Kennedy Space Center in Florida, I stood next to a group of scientists, including Dean Pesnell, SDO project scientist from Goddard Space Flight Center, as we anxiously waited for SDO's launch on board an Atlas rocket. Huddled in jackets on a sunny but uncommonly chilly Florida morning, Pesnell and colleagues talked about their "little" spacecraft tucked safely inside the nose cone of the rocket.

SDO measures fifteen feet (4.5 m) high by over six feet (2 m) wide and would be launched into what is called an inclined geosynchronous orbit about 22,000 miles (36,000 km) above Earth. There, the spacecraft can take continuous observations of the Sun while also being in an

orbit that matches the spin of Earth. That allows SDO to stay over the longitude of one location on our planet, the data-receiving center in White Sands, New Mexico. Since SDO generates huge amounts of data—about 50 times more science data than any other mission in NASA history—the spacecraft doesn't store any of it but instead sends it constantly to a single dedicated ground station.

"SDO has several improvements from previous missions," Pesnell said. "First, we have better cameras. These cameras are about twice as big as the best camera we have in space, and we have six. Our cameras are 4k by 4k pixels, so that's sixteen million pixels sitting on each camera."

He explained why seeing the full disk of the Sun in high resolution is such an upgrade. "If you see something happen, you say, 'Oh, that's cool, that's what I want to study,' and you can zoom in and look in more detail or you can zoom out to get the big picture."

But the most important thing, Pesnell said, is that SDO is going to take pictures frequently, very frequently.

"Before SDO, the best any mission has done so far is to take full disk images of the Sun almost every three minutes, and we're going to take them every ten seconds," he said. "With that, we'll be able to see things that happen on the Sun very quickly that right now we miss. It's like we've been looking at tops of the mountain and missing the valleys in between."

SDO is also studying the amount of energy and types of light our eyes can't see that is emitted by the Sun, such as extreme ultraviolet wavelengths. "These are very short wavelengths," Pesnell said, "that get absorbed very high up in our atmosphere and cause the atmosphere to heat and expand, which can bring satellites out of orbit. Previous missions have looked at the extreme ultraviolet irradiance every 90 minutes; we're going to look at it every 10 seconds."

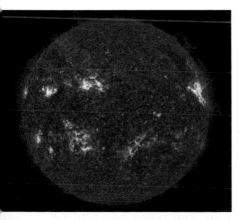

A full-disk view of the Sun from SDO in May 2012, where the Sun sported numerous prominences along its edge. The Sun is viewed here in extreme ultraviolet light. Credit: NASA/ Goddard Space Flight Center

The SDO Spacecraft

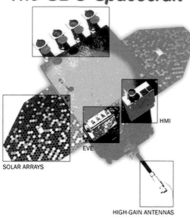

Diagram of the Solar Dynamics Observatory (SDO) with instrument locations highlighted. Credit: NASA

HMI

EVE

SOLAR ARRAYS

HIGH-GAIN ANTENNAS

The enigmatic magnetic field is also studied. "The goal is to understand the life cycle of the magnetic field," Pesnell said. "The Sun's magnetic field makes all the activity happen, and we want to know where it comes from, how it gets to the surface and then how it gets converted into solar activity."

Pesnell summed up the entire mission with a smile: "We're going to see a lot of new stuff and we're going to learn a lot."

The three instruments on SDO are designed to work together. One measures the magnetic field, one measures what the magnetic field does and the third measures what we see that affects us here on Earth. The instruments are as follows:

The **Helioseismic and Magnetic Imager** (HMI) maps solar magnetic fields and looks beneath the Sun's opaque surface using the equivalent of an ultrasound here on Earth. The instrument deciphers the physics of the Sun's activity by showing us where the magnetic field is coming from and what it looks like at the surface.

SDO launches from Launch Complex 41 at Cape Canaveral on February 11, 2010. Credit: NASA

When the Solar Dynamic Observatory (bright streak in lower right quadrant of photo) lifted off from Cape Canaveral on February 11, 2010, its launch enabled optics experts to discover a new form of ice halo. Credit: NASA/Goddard/Anne Koslosky

Project Scientist Dean Pesnell describing the launch. Credit: Nancy Atkinson

The **Atmospheric Imaging Assembly** (AIA) is a group of four telescopes designed to photograph the Sun's surface and atmosphere. The instrument covers ten different wavelength bands, or colors, selected to reveal key aspects of solar activity. These types of images are showing details never seen before by scientists.

The **Extreme Ultraviolet Variability Experiment** (EVE) measures fluctuations in the Sun's radiant emissions. These ultraviolet emissions have a direct and powerful effect on Earth's upper atmosphere—heating it, puffing it up and breaking apart atoms and molecules. EVE will help researchers understand how fast the Sun can vary at these wavelengths.

SDO launched successfully—and beautifully—that February day in 2010. And it turns out that just as the flight began, the SDO mission made a new discovery about Earth's atmosphere. Amazingly, viewers at the launchpad saw the Atlas rocket fly through a **sundog**—a bright spot in the sky created by refraction of sunlight through ice crystals found in high cirrus clouds. Shock waves from vibrations of the rocket rippled through the clouds and destroyed the alignment of the ice crystals. This extinguished the sundog and created a ripple effect around the spacecraft. Also, a bright column of white light appeared next to the Atlas and followed the rocket up into the sky. The crowd watching the launch oohed and aahed at the sight.

"We saw this sundog come out, and SDO flew right through it. Then the sundog disappeared," Pesnell said after the launch. "This may be the first time we've sent a probe through a sundog, and people will be studying this, so already we are learning things about our atmosphere from SDO."

After experts reviewed the footage of the launch, they realized that, somehow, shock waves from the rocket reorganized the ice crystals to produce the rocket halo. They had never seen anything like it and in studying the event learned about a new way for sundogs to form.

LET'S DO SCIENCE!

Six years later—to the day—I visited Pesnell at SDO's Mission Operation Center at Goddard, near Baltimore, Maryland. Except for the whir of computers and cooling fans, the room was quiet. Just one engineer checked the array of monitors. "We've been doing this for six years now,"

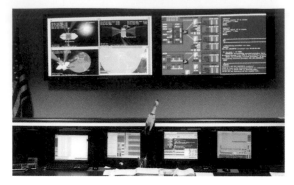

The view of SDO's Mission Operation Center at Goddard Space Flight Center in Greenbelt Maryland. A rubber chicken named Camilla watches over the mission operations team. Credit: Nancy Atkinson

Pesnell said, "and the spacecraft is so reliable and steady, on a day-to-day basis it doesn't require a big team."

Big screens showed SDO's position in space and displayed the spacecraft's field of view: a steady square bull's-eye around the Sun. A slideshow of gorgeous SDO images flashed on another large screen that Pesnell had set up for my visit. But, he confided, he had actually compiled the slideshow for a six-year anniversary celebration later that day.

Since its launch, SDO has been churning out data and images. As one SDO scientist put it, "The amount of data we have to work with is outrageously huge."

All the data, which amounts to about 98 percent of all the solar data that has ever been measured in space, is stored at the Joint Science Operations Center (JSOC) at Stanford University in California. "At JSOC, we store 24 times as much data than everybody else, combined," Pesnell said.

"The data from SDO has been eye opening," said heliophysicist C. Alex Young, who also works at Goddard. "Being able to see the full disk of the Sun in numerous wavelengths simultaneously has brought a completely different perspective. You really can't compare SDO to another solar observatory because it's like nothing we've had before."

There's a continuous firehose of data coming from SDO, Young said.

One of the first images taken by SDO, the March 30, 2010 'First Light' prominence eruption captured just after the AIA sensors were activated, shown here in the ultraviolet part of the spectrum. Credit: NASA/AIA/Goddard Space Flight Center

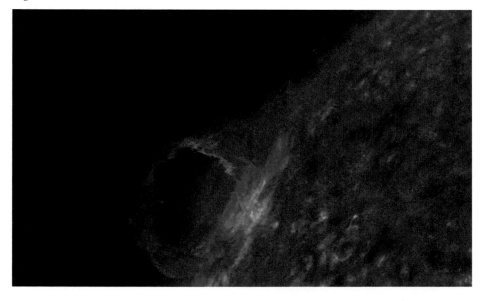

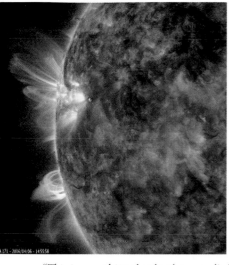

Arches of magnetic field lines towered over the edge of the Sun as a pair of active regions began to rotate into view in April 2016. Charged particles spiraling along the field lines make the lines visible when viewed in this wavelength of extreme ultraviolet light. Active regions are intense areas of competing magnetic forces that are embedded below the Sun's surface. Credit: Solar Dynamics Observatory, NASA

In six years of operation, SDO has already contributed to several groundbreaking discoveries. One of the first discoveries confirmed that the Sun has sympathetic solar flares, where simultaneous flares occur across wide distances, but in some way they are related.

"This is something that has been studied and debated for decades," said Young, "and before SDO, we didn't have enough information to really say with a great amount of certainty these near-synchronous explosions were connected."

Some scientists thought they were too far apart—sometimes millions of kilometers distant—to be related, but others felt there must be an underlying physical connection between these distant regions on the Sun.

Very early in the mission, SDO scientists had the chance to study sympathetic eruptions. On August 1, 2010, the entire visible northern hemisphere of the Sun erupted, and in a matter of hours various flares and filament eruptions occurred. SDO data showed the magnetic field lines between the flares and other events were connected. Since then, with SDO's continuous high-resolution and multiwavelength observations, scientists have seen numerous sympathetic flares connect across great distances through looping lines of the Sun's magnetic field.

Curling "surf" waves have been spotted traveling through the Sun's atmosphere in this image from SDO captured on April 8, 2010. Credit: NASA/Danny Ratcliffe, Goddard Space Visualization Studio

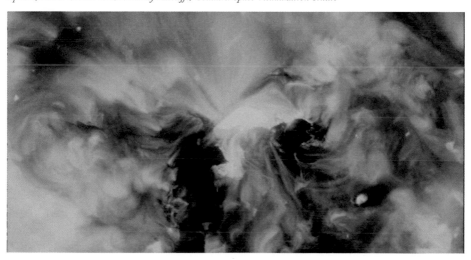

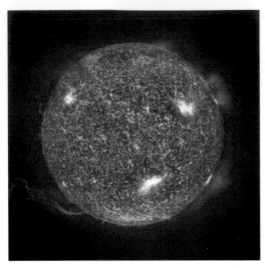

Another finding from SDO is the first-ever full observations of super-high-speed **solar waves**—sometimes called coronal waves or solar tsunamis—where a wave of hot plasma "surfs" along the Sun's surface.

"We've known about coronal waves since shortly after the Solar and Heliospheric Observatory (SOHO) mission launched in 1995," said Young, "but it was difficult to see them in their entirety because of the small field of view we had with other observatories. Now, we can really see them traveling across the Sun, we can see where they pass an active region how the wave interacts with them."

Young said it's like watching waves on a pond that encounter obstacles like rocks and observing how the waves respond to the rocks.

Pesnell agreed that the most important advance provided by SDO is being able to have a detailed global view of the Sun.

"We see the entire Sun at once and can see how even initially small events like the solar waves can move and cause other things to happen," he said. "We've been tracking these waves, and by watching how the waves propagate and bounce off things, we're learning more about the Sun's lower atmosphere, which is helping us to diagnose and begin to predict what will happen next."

SDO also has brought scientists closer to solving what is considered the most intriguing enigma in modern solar physics: the coronal heating mechanism and why the corona is significantly hotter than the Sun's surface, the photosphere.

"If you're standing next to a fireplace," Pesnell said, "and you move away, you begin to feel cooler. But that's not how it works on the Sun. On this otherwise normal, rather boring-looking star, why do we have a corona sitting on top that is about 200 times hotter?"

It's as if the air around this scorching, burning fireplace was hotter than the fireplace itself. And if the corona is so hot, why doesn't it heat up the Sun's surface to a similar temperature?

While the mystery still isn't fully solved, astronomers now have a better understanding of the mechanisms of coronal heat generation and transfer. The leading theory is called **nanoflares**. While nanoflares are a billion times less energetic than ordinary flares, Pesnell said, these types of flares are going off on the Sun almost nonstop.

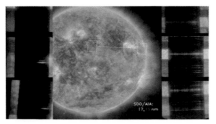

A NASA sounding rocket examined light from the Sun in the area shown by the white line (imposed over an image of the Sun from NASA's Solar Dynamics Observatory) and then separated the light into various wavelengths, as shown in the lined images—spectra—on the right and left, to identify the temperature of material observed on the sun. The spectra provided evidence to explain why the Sun's atmosphere is so much hotter than its surface. Photo courtesy of NASA/EUNIS/SDO

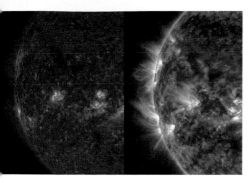

This side-by-side rendering of the Sun at the same time in December 2015 show two different wavelengths of extreme ultraviolet light. These different views help scientists to visualize the differing features visible in each wavelength. Here, finer strands of plasma looping above the surface in the 171 Angstrom wavelength (gold) than in the 304 Angstrom wavelength (red), which captures cooler plasma closer to the Sun's surface. SDO observes the Sun in 10 different wavelengths with each one capturing somewhat different features at various temperatures and elevations above the Sun. Credit: Solar Dynamics Observatory, NASA

"Basically, these are tiny flare-like events that happen almost constantly near the surface of the Sun," he explained. "And we think they are like heating elements."

Think of heating elements that run through an electric blanket. Just as one small element can't heat the entire blanket, individual nanoflares can't heat the entire corona. But together, the tiny but consistent flares send enough energy to heat up the entire blanket of the solar atmosphere.

"It's a great feeling knowing that SDO data is helping us to better understand this long-standing question," Pesnell said, "Nanoflares seem to fit what we're seeing in both SDO data and from other satellites looking at the Sun, too."

SDO has also overturned previous notions of how the Sun's writhing insides move from equator to pole and back again, providing a key part of the understanding of how the physical process that generates the Sun's magnetic field called the **dynamo**—works. Modeling this system also lies at the heart of improving predictions of solar events and the intensity of the next solar cycle.

Pesnell said the goal is to someday be able to predict solar storms with great accuracy.

"With SDO, when something big happens on the Sun, we can go back and watch what happened prior, to see what may have caused a big eruption or flare," he said. "What we are looking for is a predictor in the magnetic field lines. Right now, we can watch the magnetic fields in the active regions rearrange, and we've been hoping that there would be a regular, predictable pattern. But unfortunately it hasn't quite worked like that. Not all events are the same, and we haven't gotten quite sophisticated enough yet to fully understand how the magnetic field works to make these predictions."

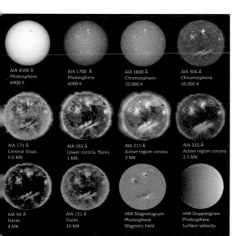

Can you see the "Man in the Sun?""Features on the Sun create a face, but it's not visible in every wavelength. 12 different views of the Sun in various wavelengths, taken at the same time with two instruments on SDO, the HMI and AIA. Credit: NASA/SDO/Goddard Space Flight Center

But that's just how the scientific process works, Pesnell said. "People try a theory and follow it until it either continues working or it breaks," he said. "It can be frustrating, but it's also a lot of fun."

However, SDO provides a window into how the magnetic fields change over time, helping scientists figure out what triggers giant solar flares and CMEs. Perhaps one day, just as meteorologists can predict weather on Earth, solar physicists will be able to reliably predict solar storms, thanks to SDO.

WE <3 SDO

During the early part of the mission, there was an unusual but heartfelt connection between the complicated SDO mission and the general public. This connection came from a rubber chicken named Camilla.

The story of Camilla goes back to the early days of SDO at Goddard Space Flight Center, well before the mission launched. Project scientist Barbara Thompson introduced Camilla as a fun diversion for the science team at Goddard, with the only obvious connection being that a rubber chicken is the same color as the Sun. Initially Camilla was just an interoffice morale booster and team builder, but over time she became more and more integrated into the education and public outreach side of the mission.

Pesnell and Young both said Camilla made a big difference in how they could interact with the public, especially with children.

"We did a lot of educational activities," Pesnell said, "and our goal was always to engage people by getting them to ask us questions. But it seemed that I—a bearded PhD scientist—kind of intimidated kids, and they felt they couldn't ask me anything. But as soon as you brought in Camilla, all the hands went up."

Camilla's official "bodyguard" was Romeo Durscher, who worked with the SDO team at Stanford University. "We always thought Romeo did a great job of making Camilla accessible to a wide audience, and that was an important part of the educational aspect of the mission," Young said.

NASA has since changed the process of education and public outreach, and Camilla is no longer the official mission mascot. But Camilla Corona (she is going by her full name now) is still active through social media channels to encourage students to pursue studies and careers in STEM (science, technology, engineering and math) fields. Additionally, the flight operations team has re-adopted Camilla, and she watches over the SDO MOC at Goddard.

"She's now just in the background," said Pesnell, "but Camilla was great for outreach and probably the best example of where people took a shine to a mascot and as a result grew attached to a satellite and its images."

The interest and affection for SDO's incredible images—initially fueled by Camilla—has been enhanced by an easy-to-use online tool called the Helioviewer.

"Well before SDO launched," said Young, "one of my colleagues, Dr. Jack Ireland, was considering how we were going to be able to look at these large 4k by 4k images and be able to zoom in and navigate through this vast amount of solar data."

Ireland was inspired by the technology in Google Maps and other sites using large data sets in online browsers. He came up with the Helioviewer, which has not only been a critical tool for the scientists, but it's also been available to the public, giving everyone access to all the incredible views of the Sun.

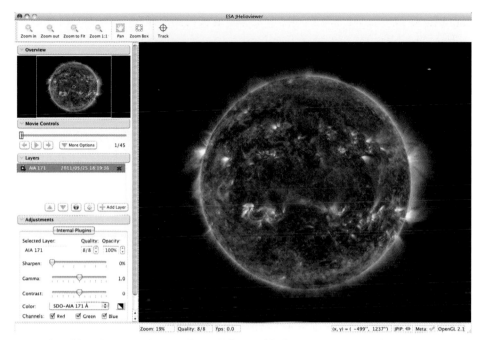

A screenshot of the Helioviewer website. Credit: Geeked On Goddard

Helioviewer is a desktop program where users can search through images of the Sun from SDO and data from several other solar spacecraft since 1991 as well as watch real time current data. Users can create movies, overlay different images and wavelengths, zoom in and out, and follow a solar event from start to finish. The movies are easily shared, with the option to upload them directly to YouTube. So far, millions of solar movies have been created on Helioviewer.

The Helioviewer site also has an online space for Sun watchers to gather and chat and has become a great example of crowdsourcing. Citizen scientists using Helioviewer collaborate on movies, and sometimes they even alert the "real" scientists to events happening on the Sun.

Young recalled the morning of June 7, 2011. He had just sat down with his morning cup of coffee at home, when he got an email from Jack Ireland.

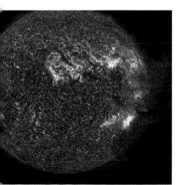

The huge fountain of solar material erupting from the Sun on June 7, 2011. Credit: NASA/SDO/Goddard Space Flight Center

"Jack told me I had to take a look at a video a user had created," Young said, "and the video showed a gigantic fountain of solar material that had just exploded off of the Sun. I had never seen material released like this before. It looks like someone just kicked a giant clod of dirt into the air and it fell back down on the Sun."

Young said that event was a big turning point for him. He created his own narrated video about the event, which launched his own love of interacting with the public about the Sun.

"I really enjoy science communication and public speaking and sharing all this cool stuff," he said excitedly. "Sharing results and interacting with students—young and old—gives you little jolts of adrenaline and excitement to continue forward."

NO "KILLER" SOLAR FLARES

You might still be thinking about the near miss, the mega solar storm that occurred in 2012 described at the beginning of this chapter. Or you may have heard doomsday predictions about "killer" solar flares that could destroy our atmosphere, wiping out life on Earth. We know that solar flares and CMEs happen all the time on the Sun, and they hit Earth once or twice a week, sometimes more. Should we be worried?

Yes and no. For the most part, no.

"Our Sun is an active star, and it goes through a cycle of activity every eleven years, and this has been happening throughout history, as far as we can tell," said Young. "We always have solar flares; sometimes they are big, sometimes they are small. But even in the largest events, we've seen that, in general, the effects are minimal."

In recorded history, we only know of a few big events where solar storms caused problems on Earth:

- September 2, 1859: The largest solar storm ever recorded disrupted telegraph service.

- March 9, 1989: A CME ejected from the Sun, and when it reached Earth on March 13, it caused the collapse of the power network across Quebec, Canada. A blackout lasted more than eleven hours and affected over six million people. The United States experienced over 200 power grid problems but no complete blackouts. Some aircraft and satellites experienced communication interruptions.

- December 5 and 6, 2006: X-class flares on the Sun triggered a CME that interfered with GPS signals being sent to ground-based receivers and radio transmissions to and from aircraft.

Today, airlines fly over 7,500 polar routes per year. Certain space weather events can cause radio blackouts in the north polar region. These events can last for several days, during which time aircraft must be diverted to latitudes where satellite communications can be used.

Solar storms pose a risk mainly to the technology we've developed such as electrical grids and satellites. Young explained that Earth's relatively thick atmosphere and our planet's magnetosphere stop all the harmful radiation that is produced in a solar flare and CME, and so humans are not in any danger. In even the largest events, no solar storm would strip away Earth's atmosphere.

"You just can't get that much energy out of the Sun," Young said. "It would have to go supernova, and our Sun is the type of star that just won't do that."

Larger solar storms can cause auroras in the northern latitudes, which are actually a beautiful, eye-catching benefit. But, yes, these geomagnetic storms can adversely affect satellites and power grids.

"But these are the kinds of things that people who operate these systems know about and have learned to prepare for," Young said. If a massive solar storm were going to hit Earth, power grids could be temporarily shut down and satellites could be put into a safe mode until the storm passed. NASA has actually been working with power companies and satellite operators to help develop standards and guidelines to keep power grids stable during geomagnetic storms.

And we have warning. CMEs travel at speeds that take them anywhere from 12 to 36 hours to reach Earth.

"We've learned more and more about storms," Young said. "We've learned how to better predict the effects and where they are going to hit. As long as we continue paying attention and learning more about them and treat them like a hurricane or huge thunderstorm, we can take appropriate measures and be ready for them."

Beautiful aurora results from charged particles from the Sun interacting with Earth's atmosphere, such as this stunning view from northern Minnesota. Credit: Bob King

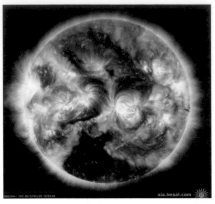

A large, dark coronal hole at the bottom of the Sun in January, 2014. Coronal holes are areas where the Sun's magnetic field is open ended and where high-speed solar wind streams into space. The area appears darker there because there is less material being imaged in this combination of threes wavelength of extreme ultraviolet. At its widest point, the hole extends about half way across of the Sun, close to 50 times the size of Earth. Credit: Solar Dynamics Observatory, NASA

We know we can't stop solar storms, but we can prepare for them.

Young added that we "understand the Sun well enough now with spacecraft like the Solar Dynamics Observatory monitoring 24 hours a day, 7 days a week to know that any type of super solar storm that might wipe out Earth simply isn't going to happen. It is a physical impossibility."

SDO AND THE FUTURE

Pesnell said the SDO spacecraft remains healthy, and the instruments maintain the same accuracy as they did right after launch. He expressed his appreciation for the team of engineers and scientists that built such a robust spacecraft, knowing SDO could be a long-term mission.

"We just have to keep SDO pointed at the Sun," he said, "but we have plenty of fuel, enough to keep going for decades if the basic infrastructure of the spacecraft holds out. But in the meantime, we're just going to keep the data flowing."

Young agreed, saying that SDO has become a critical tool not only for solar scientists, but also for the broader space science community.

"I think the community has now gotten so used to having all this wonderful imagery and detailed data all the time," he said. "I look at the images every day and there is always something interesting, new or surprising. The imagery just blows me away all the time."

CHAPTER 8

RISING TO THE OCCASION: THE MARS RECONNAISSANCE ORBITER AND HIRISE

RISING FROM THE ASHES

Within three months, two NASA missions to Mars failed in dramatic fashion. The Mars Climate Orbiter vanished just as it reached the red planet on September 23, 1999. Then on December 3 of that year, the Mars Polar Lander crashed during its landing attempt.

An artist's rendering of the Mars Reconnaissance Orbiter around Mars. Credit: NASA/JPL-Caltech

NASA's Jet Propulsion Laboratory sits among the foothills of the San Gabriel Mountains in Pasadena, California.
Credit: NASA/JPL

Ensuing investigations revealed the orbiter likely entered Mars' atmosphere and burned up. The reason? There was a mix-up between metric and English units of measure used in ground-based software to monitor the spacecraft.

The lander's crash, it was later determined, was due to an errant signal sent to the computerized flight-control system. When the landing legs deployed, the computer thought the spacecraft had landed when, in fact, it was still 130 feet (40 m) up. The landing rockets shut down prematurely and the probe plummeted to the surface.

The losses spurred several intense investigations, both internal and external to NASA. Reports from the inquiries revealed the mission teams suffered from understaffing, inadequate safeguard procedures and communication problems. The two mission failures and a previous failure in 1993 compelled NASA to review their entire Mars program.

With these losses, the turn of the millennium was a tough period of time to be working at NASA—in particular at JPL, the home of these missions. One JPL official, Firouz Naderi, who was tasked with revamping NASA's Mars exploration program in the aftermath, commented that if there were such a thing as institutional depression, JPL had it.

This was the culmination of the NASA era of "faster, better cheaper," a mantra coined by then-NASA administrator Dan Goldin, who wanted NASA to cut costs while still delivering a wide variety of missions. While this approach made it possible for missions like New Horizons and Kepler to be built and launched, sometimes cutting costs ended up meaning cutting corners.

"We knew cutting corners meant we could possibly lose a spacecraft," said Rich Zurek, who was the project scientist for both missions. "But losing two or three? That was not acceptable."

The failures prompted a reevaluation and a move toward a more reasoned and balanced approach to space exploration.

"We needed to approach it as each one of these spacecraft is precious, each of these we want to work," Zurek said. "So let's put our best effort into it, always. And that's now paid off in terms of mission success and longevity of the spacecraft."

To rebuild the ailing Mars program and boost morale, just a year after the losses NASA announced a new strategic plan for exploring Mars with a series of orbiters, landers and rovers.

Rich Zurek, left and HiRISE Principal Investigator Alfred McEwen, right, at a press conference at Kennedy Space Center prior to MRO's launch. Credit: NASA/JPL

The backbone of the program would be a new and much larger Mars orbiter that would carry out a survey of potential landing sites for future missions, with the ultimate goal of humans landing there one day.

That orbiter, the Mars Reconnaissance Orbiter (MRO) has now been circling Mars for over a decade. MRO is an example of the long-lived missions that have become the hallmark of Mars exploration since the turn of the millennium, as NASA has now established a remarkable record of success, with a string of seven successful orbiters and landers conducting comprehensive investigations of the red planet. And there are more missions on the way.

Zurek is now the head of NASA's Mars exploration program as well as the project scientist for MRO.

Along with rovers Spirit, Opportunity and Curiosity, as well as other orbiters, including NASA's Mars Odyssey and ESA's Mars Express, MRO represents the new wave of robust spacecraft conducting prolonged missions. These extended missions provide a constancy that gives researchers the data they need to study seasonal and longer-term processes on the red planet.

"Because of enduring missions like MRO, the rovers and other orbiters, we now know that Mars is an incredibly complex and diverse place," Zurek said. "It's an active world, with changes taking place today."

COMPETITION

Competition is a big part of spaceflight. Competition exists between missions to win funding, between science teams to have their instruments chosen and there are even competitions to decide where to set down a rover or lander.

MRO emerged from a competition and actually lost, if only briefly.

"After the two missions failed, the question was what should our next steps be?" Zurek recalled. "Studies were commissioned, and the leading candidates were to create a rover, and another was for an orbiter to do reconnaissance for future landing sites. These two concepts were in competition with each other, and finally it was decided to go with a rover. I believe they thought wheels on the ground were more exciting. But the NASA administrator, Dan Goldin, proposed two rovers. In some respects, that was because we just lost two spacecraft, so let's make sure at least one works."

That led to Spirit and Opportunity's missions, which launched toward Mars on June 10 and July 7, 2003. Zurek said they were quite disappointed when the rovers went first, leaving the orbiter in limbo for a time, but everyone knew the orbiter was needed.

"The orbiter would look for the best landing sites, to 'follow the water,' which has been our theme in Mars exploration," Zurek said, "and see things at a scale that could make a difference, assessing both the safety and the science for future discoveries, with the grand themes of looking at water, climate change, the potential for life."

In building MRO and preparing for the mission, there was the pressure to succeed, said Dan Johnston, who is the program manager for MRO. He said several new processes came into play during MRO's development as NASA moved from faster, better, cheaper to establishing better practices. But it wasn't always easy.

"If you think of driving a car down the road," Johnston said, "perhaps we went too far to one side and went off the road in the missions that were lost. The challenge was to not be too conservative so that you go off the road in the other direction, but to drive on the road in the right way, down the middle. That was part of the challenge of the environment at that time."

Being able to work under difficult constraints took a unique mix of engineers, many with alpha-type personalities who had a drive and passion to turn things around for NASA. Johnston recalled one person telling him it was the first time she had seen so many alphas working together and not constantly being at each other's throats.

But, Johnston said, the team formed to build a next-generation orbiter was excellent. And all the team members had that competitive drive.

"Jim Graf was the project manager and he put together a very strong team for mission design and operations that worked together very well," Johnston said. "We were very competitive with each other and I think in some regard quite aggressive. However, Jim Graf's leadership gave us the ability to work well together and be committed."

And they had fun together, too.

"There was a nearby Mexican restaurant that became our venue for after-work and special events," he said. "Spending time together away from work broke down some of the stress barriers, and we had developed good relationships. I think that helped with the success of the mission."

But really, the competition never ends.

"There has always been competition between all the Mars missions, with some saying that if you're not a lander or a rover, you're not anything," Johnston smiled. "The Opportunity team likes to say they've done a marathon across Mars' surface, and our orbiter team comes back with, 'We cover that in ten seconds.'"

And with all due respect to the other orbiters, Johnston says, nothing is like MRO. "This spacecraft's capability is very impressive."

"These are all very talented people and competition is part of the game," Zurek said, "to do it better, to do it well, to take on challenges that you don't know how to solve at the beginning. Part of my enjoyment of this mission is being able to work with experts from all over the world. Any topic you need to know about, you can find someone who is an expert in it, bring their expertise to bear on the problem and make progress."

MARS RECONNAISSANCE

NASA wanted a high-caliber, robust orbiter that could obtain a high-resolution global view of what was happening on Mars today while also gathering clues to help tell the story of Mars' mysterious past. Did water really flow there long ago—or today? Was there ever life on Mars? If so, what happened to it, and could those same processes take place on Earth?

"From the beginning, the intent for MRO was to send a very capable spacecraft to support not only fundamental science, but to support the Mars program in the long term," said Johnston.

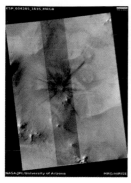

A dramatic, fresh impact crater dominates this image taken by the High Resolution Imaging Science Experiment (HiRISE) camera on NASA's Mars Reconnaissance Orbiter on Nov. 19, 2013. HiRISE team member Kristin Block calls this the "tie-dye" crater. Credit: NASA/JPL-Caltech/Univ. of Arizona

Dark, narrow streaks on Martian slopes such as these at Hale Crater are inferred to be formed by seasonal flow of water on contemporary Mars. The streaks, called "recurring slope lineae" or RSL, are roughly the length of a football field. The imaging and topographical information in this processed, false-color view come from HiRISE. Credit: NASA/JPL-Caltech/Univ. of Arizona

"The concept was to develop an orbiter that could deliver reconnaissance capability to be able to image specific targets on the surface, down to one-meter resolution."

That means objects as small as three feet (1 m) across would be easily visible from orbit. MRO's other job was to be carry a telecom package to provide relay communications for current and future landers and rovers on Mars' surface.

NASA's Mars exploration theme of follow the water directed the MRO mission goals: search for sites showing evidence of water-related activity, characterize the present climate of Mars and its seasonal change, determine the nature of complex layered terrain and identify sites with the highest potential for future landers and rovers.

The suite of instruments would provide the ability to zoom in for extreme close-up photography of the Martian surface, to analyze minerals, look for subsurface water, trace how much dust and water are distributed in the atmosphere and monitor daily global weather.

Since the opportunity to go to Mars comes about every 26 months—when Mars and Earth reach a position in their respective orbits that offers the best trajectory for the quickest travel time between the two planets—the MRO team knew they had to make their launch window in the summer of 2005.

On August 12, 2005, MRO lifted off from Cape Canaveral Air Force Station and seven months later, the orbiter arrived at Mars, beginning an incredible journey of exploration.

Telescopic Camera for MRO. The High Resolution Imaging Science Experiment (HiRISE) is one of six science instruments for NASA's Mars Reconnaissance Orbiter. Credit: NASA/JPL

HiRISE

"We've had a very dependable spacecraft with some great instruments," said Johnston.

The Context Camera (CTX) captures wide-area images to provide context for the high-resolution images, while the Mars Color Imager (MARCI) images weather features such as clouds and dust storms. The Compact Reconnaissance Imaging Spectrometer for Mars (CRISM) is used to identify minerals on the surface. The Mars Climate Sounder observes Mars atmosphere, whereas the Shallow Subsurface Radar (SHARAD) looks for signs of water ice below the Martian surface.

"The spectrometer detecting the different mineral deposits is wonderful," said Johnston, "but with all due respect to the other science investigations, I think that imagery with HiRISE is of a completely different class than anything else."

HiRISE, the High Resolution Imaging Science Experiment, is the largest and most powerful camera ever flown on a planetary mission. While previous cameras on other Mars orbiters can identify objects about the size of a school bus, HiRISE brings it down to human scale.

"HiRISE provides very high-resolution imagery, showing Mars at the scale of people, one-meter resolution," said Alfred McEwen, the principal investigator for HiRISE. "We see features that you can relate to if you were there on the surface. You can imagine yourself hiking around over the surface as being seen by HiRISE."

This artist's concept represents the "Follow the Water" theme of NASA's Mars Reconnaissance Orbiter mission. The orbiter's science instruments monitor the present water cycle in the Mars atmosphere and the associated deposition and sublimation of water ice on the surface, while probing the subsurface to search for water-ice. Credit: NASA/JPL

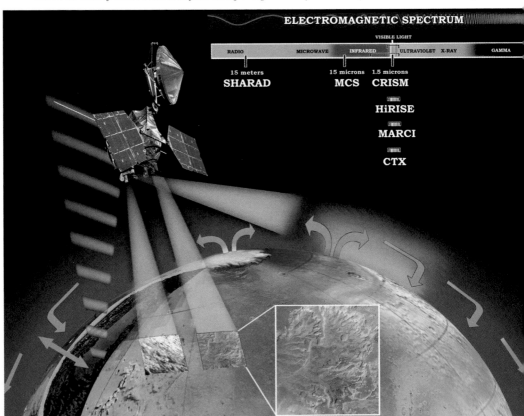

That capability has allowed the orbiter to identify obstacles such as large rocks that could jeopardize the safety of landers and rovers.

Even before the mission, McEwen had high hopes—and high expectations—for his camera. "The very high-resolution imaging was seen as essential to the effort of doing reconnaissance," he said, "and while we weren't required to have color or stereo (3-D) color imaging, we wanted to build those extra capabilities into the camera, as I wanted them for the direct science potential. And now color and stereo are seen as essential to characterize future landing sites."

Johnston said HiRISE allows MRO to be a much more powerful microscope on Mars than was ever imagined. "Even in looking at next-generation orbiters, they'd have to go a long way to beat the capability of this spacecraft, particularly in its landing site scout capability," he said.

HiRISE does what is called **landing site characterization** to help determine landing sites for future missions. This includes not only NASA missions, but missions from other space agencies as well.

"HiRISE can characterize an area by taking images over multiple passes of potential sites," explained Johnston, "and the science teams certify the sites as safe or not by looking at slopes, boulders and other hazards. For all future missions, we have a process in place where mission teams from anywhere in the world can submit requests to us and we try to schedule those requests as targets for study. It's a great example of international cooperation of the missions."

HiRISE is equivalent to a 20-inch (51 cm) telescope with state-of-the-art digital detectors. With the spacecraft orbiting at just 190 miles (300 km) above Mars, it has now taken hundreds of thousands of close-up images of the red planet that contain scientifically useful information as well as incredibly beautiful views of Mars.

"Mars' surface is covered by bright dust," said Zurek, "and so it tends to look the same everywhere. It's a bit like going to the desert here on Earth, and at first glance everything seems the same. But as your eyes get adjusted to things, you see more, a lot more."

Zurek said that over time they can see changes taking place in locations that are imaged several times, seeing things like new craters and sand dunes that move. But processing HiRISE images in a special way allows the colors to be "stretched"—making the blues really blue, the reds extra red—to bring out subtle highlights and important features.

MRO took this image of the complex, layered sedimentary rocks on the floor of an impact crater north of Eberswalde Crater. There may have been a lake in this crater billions of years ago, and the area was once considered for a landing spot for the Mars Science Laboratory. Places such as Eberswalde Crater, where sedimentary rocks may have been laid down by standing water, are ideal locations to search for evidence of past microbial life if it ever developed on Mars. Credit: NASA/JPL-Caltech/ University of Arizona

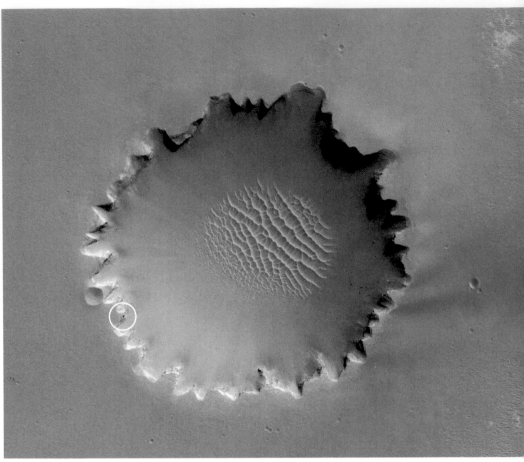

Victoria Crater at Meridiani Planum. A close look at the 8 o'clock position of the crater reveals the tiny Opportunity Rover. Credit: NASA/JPL-Caltech/University of Arizona/Cornell/Ohio State University

HiRISE has also taken some very unique and unusual images. Just seven months after its arrival, HiRISE peered down from orbit and snapped a picture of a crater called Victoria. There sat the Opportunity rover, perched on the crater's rim. Not only was the rover visible, but also the tracks created by its wheels showed the path the rover had traveled. Since then, HiRISE has captured images of the Spirit rover, the Phoenix lander, the Curiosity rover and has even found evidence of crashed landers, from previous failed missions.

Incredibly, HiRISE has caught avalanches and dust storms in action on Mars. The avalanches are a result of carbon dioxide frost that clings to steep slopes in the darkness of winter, and when sunlight hits them in the spring they loosen up and fall.

McEwen said these events happen mostly in the middle of spring, roughly equivalent to April to early May on Earth, and since HiRISE has captured events like this several times, it is now known to be a regular spring process at Mars' north pole region every year.

HiRISE also captured a large **dust devil**—a tornado-like serpentine dust storm—scooting across Mars' surface. These dust plumes can reach more than half a mile (800 m) in height and about 30 yards or meters in diameter.

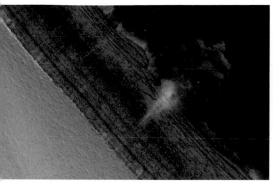

Caught in the act by the Mars Reconnaissance Orbiter: an avalanche cascades down a steep, icy scarp near the north pole of Mars, kicking up a cloud of reddish dust about 200 meters wide. NASA /JPL / University of Arizona

A towering dust devil casts a shadow over the Martian surface in this image acquired by HiRISE. Credit: NASA/ JPL-Caltech/Univ. of Arizona

Perhaps the most incredible images taken by HiRISE are of two other spacecraft. These two images are unique in that the HiRISE captured the images of these spacecraft *as they were descending to Mars on a parachute!*

For Phoenix, which landed in May of 2008, it was the first time a spacecraft was imaged during the final descent onto a planetary body. HiRISE was pointed toward the area of Phoenix's descent, and from a distance of about 472 miles (760 km) away, it captured Phoenix with its parachute descending through the Martian atmosphere. The image reveals the 30-foot (10-m)-wide parachute fully inflated.

"For Phoenix we got a bit lucky with HiRISE in terms of the geometry, giving us a high probability of success," McEwen said.

For the Curiosity rover landing in August of 2012, the HiRISE team decided to attempt a repeat performance. How difficult was it?

MRO's HiRISE camera acquired this dramatic oblique image of Phoenix descending on its parachute. Shown here is a wider view of the 10 kilometer (6 miles) diameter crater informally called "Heimdall" and an improved full-resolution image of the parachute and lander. Although it appears that Phoenix is descending into the crater, it is actually about 20 kilometers (12 miles) in front of the crater. Credit: NASA/JPL-Caltech/Univ. of Arizona

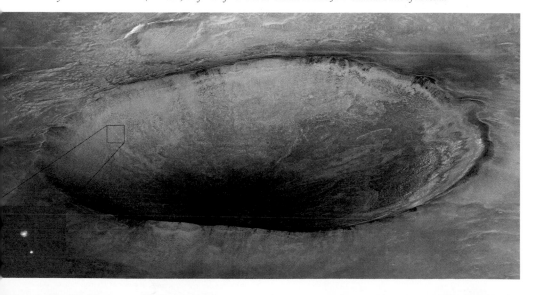

NASA's Curiosity rover and its parachute were spotted by HiRISE as Curiosity descended to the surface on August 6, 2012, while the orbiter was also listening to transmissions from the rover. Curiosity and its parachute are in the center of the white box and the rover is landing on the etched plains just north of the sand dunes that fringe Mount Sharp. MRO was 340 kilometers (210 miles) away from Curiosity when the image was taken. Credit: NASA/JPL-Caltech/ Univ. of Arizona

"If you've ever taken a picture out the window of a moving car, you know that some portions of the picture are crisp and in-focus but due to the motion of the car, others are blurred," said Sarah Milkovich, a scientist with HiRISE at the time of Curiosity's landing. "For this once-in-a-lifetime image, the HiRISE team needed to make sure that the descending rover was crisp even if the background surface was blurry. Plus they needed to ensure that the exposure time was not too long (overexposing the bright parachute) or too short (making the image too dim and noisy to see anything)."

Calculations had to be exactly correct, as there was just one chance to get it right, with only enough time to take just one image. It required coordination between MSL navigation team, the MRO navigation team and the MRO flight engineering team. The final set of commands was uplinked to MRO three days before landing, and the teams waited nervously to see if all of their calculations and predictions were accurate.

They nailed it.

"HiRISE had taken over 120 pictures of Gale Crater in preparation for Curiosity's mission, but I think this is the coolest one," Milkovich said.

HOW TO TAKE A PICTURE FROM A SPEEDING SPACECRAFT

I asked McEwen if the quality of images from HiRISE has surprised him.

"No, not at all," he said. "Prior to the mission, there were a lot of doubts expressed by members of the planetary science community, and the main issue was not how well the HiRISE camera would perform but the spacecraft stability needed to get such sharp images. Credit the engineers for doing a great job on the spacecraft."

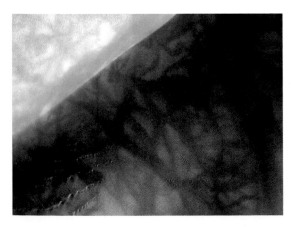

The Russell Crater dune field is covered seasonally by carbon dioxide frost, and this image shows the dune field after the frost has sublimated (evaporated directly from solid to gas). There are just a few patches left of the bright seasonal frost. The dark streaks are dust devil tracks meandering across the dunes. NASA/JPL-Caltech/University of Arizona

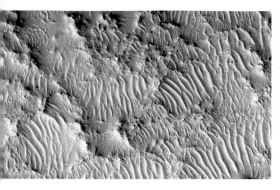

A Possible Landing Site for the 2020 Mission: Jezero Crater. It's not only when trying to find a scientifically interesting place to land that the high-resolution images from HiRISE come in handy: it's also to identify potential hazards within a landing ellipse. This is one of the trickier aspects of selecting landing sites on Mars: a place to do good science but also where the risks of landing are low. Credit: NASA/JPL-Caltech/University of Arizona

As with every space mission, the engineers and navigators who work behind the scenes are the unsung heroes enabling mission success.

"We have to provide accurate predictions of where the spacecraft is going to be," said veteran JPL spacecraft navigator Neil Mottinger, who worked with the MRO navigation team for launch and the early part of the mission. "Then the engineers know how to orient spacecraft so the scientists can make their observations. If we do our job, the scientists can see an avalanche on Mars or look at specific areas on the planet. If our predictions are wrong, the cameras are pointed in the wrong direction. Navigation is integral to the whole process of ensuring mission success."

MRO speeds around Mars at about 1.8 m a second (3 km/second). Continuing with Milkovich's analogy of taking images from a moving car, just how challenging is it to take images with a spacecraft going that fast?

"Oh, it's easy," said Christian Schaller with a smile. Schaller is the software developer responsible for the primary planning tools the HiRISE targeting specialists and science team members use to plan their images. "Really, it is easy, because a large team of incredible engineers and scientists have put together a fantastic spacecraft and a fantastic instrument, and NASA's navigation ability is top-notch. So in practice, it looks almost effortless to us on the imaging side because these incredibly talented people have hidden all the gory details from us."

Because of the intricate coordination between the imaging and navigation teams, they all know exactly where the spacecraft is at all times and exactly how fast MRO is moving.

The coordination itself is in the form of **ephemerides** (computations of exactly where planetary bodies and spacecraft are at a given time) delivered through a mechanism developed by JPL's Navigation and Ancillary Information Facility (NAIF). And that includes some seriously complicated acronyms.

"The mechanism is called SPICE, which is a meta-abbreviation that stands for SPK, PCK, IK, CK and EK (spacecraft and planetary ephemerides, planetary constants, instrument descriptions, pointing information and event information)," Schaller explained. "The MRO Nav team generates our orbit information, converts it to SPICE ephemerides files and delivers them to the extended project. Our planning tools ingest the SPICE files and show us where we'll be (predicted orbits) and where we've been (reconstructed orbits)."

Schaller said the main challenge for the HiRISE team is matching the camera's scan rate with the projected ground speed. "If our scan rate is wrong, we'll get smeared or distorted images," he said.

And that's where the software Schaller developed comes in, which is responsible for computing the scan rate based on the projected ground speed.

The first image taken by HiRISE in Mars' orbit, the Bosporos Planum Region. Credit: NASA/JPL/University of Arizona

HiRISE's first image from the science orbit, a favorite of Alfred McEwen, shows layered deposits in the floor of Ius Chasma. Credit: NASA/JPL/University of Arizona

Schaller recalled the first time his software was put to the test: the first image taken from Mars orbit by HiRISE. It was acquired during aerobraking, where the spacecraft used Mars' atmosphere to gently slow itself down into the science orbit, which took a couple of months.

"We were all pretty nervous and excited about this image," he said, "and I was more than a little apprehensive. If I'd screwed up the scan rate calculation, then I'd have had a lot more work ahead of me as the spacecraft settled into its final orbit."

The team organized an event in their planning center so they could be together when the data from that first image were coming down. "We were all blown away to see it for the first time," Schaller said.

For the most part, instead of just taking random images, Schaller said the imaging team almost always starts with a desired target of some kind: a spot on Mars that has been requested, either a request from a science team member, a request for landing site analysis or a suggestion from the public (details on that later).

"We start with a target," Schaller explained, "then we figure out whether we can observe it at all and, if so, when. We coordinate with other teams. Once we've got that worked out, we set our image parameters, generate our observation commands and deliver them to JPL, and the spacecraft engineering team sends them to the spacecraft."

For special images like the Phoenix and Curiosity landing, the coordination is even more extensive. And once the images are taken, then comes the chore of actually finding the specific target in the large, high-resolution raw images sent from the spacecraft.

"Since HiRISE is really fourteen different cameras arrayed side-by-side (with two above the main array and two below), we have fourteen different images to look through," Schaller said. "So we hunt through twenty-eight separate files, and each of those files is 1024 pixels wide by a few tens of thousands of pixels tall, with the target, like Curiosity or Phoenix is just a few pixels across."

So the hunt takes a while, and they couldn't be 100% sure they actually captured the tiny spacecraft.

"But it's very clear when someone's found it, because of the shouting that takes place in that person's office," Schaller laughed.

"We had imaged the heck out of Gale Crater in the months leading up to landing," said Kristin Block, who is a Spacecraft Science Planning Engineer for HiRISE. "That area was probably more familiar to us than any other on Mars. We combed through the images for any sign at all of something new on the landscape. There was lots of squinting, head tilting and false

HiRISE also captured Curiosity's parachute and backshell laying the surface of Mars, near the MSL landing site. Credit: NASA/JPL/University of Arizona

alarms before we scrolled to the parachute and backshell. Much of Mars is dusty, almost monochrome in places, but that parachute—it's so bright and in that image it almost looks like it's glowing."

Block said she and her colleagues were elated they were successful and that they came through for their MSL colleagues. "But mostly at that moment I was giddy seeing a human-made spacecraft descending to the surface of another planet, and I just had to touch it on the screen."

Capturing both Phoenix and Curiosity descending through Mars' atmosphere has given the HiRISE team confidence they can accomplish, to be honest, just about anything.

"I think the main lesson Phoenix and Curiosity taught us is that we actually could do this, and that, you know, humans are awesome," Schaller said. "It's one thing to look at a picture of Mars taken from orbit and marvel at the human race. It's another thing entirely to look at a picture one spacecraft took of another during a slightly chaotic 7-minute descent, where the HiRISE field of view overlaps the EDL module's trajectory for only 40 or so seconds. That's breathtaking on a whole new level."

MRO'S DISCOVERIES

Some of MRO's discoveries show the variety of things the different instruments can do. These include identifying underground geological structures, scanning atmospheric layers and observing the entire planet's weather daily. Also, MRO determined there is enough carbon dioxide ice buried in the south polar cap to double the current atmosphere if it were released in gaseous form.

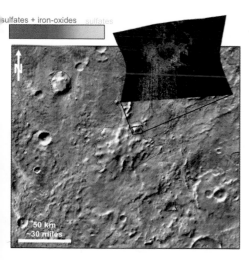

sulfates + iron-oxides   sulfates

This graphic illustrates where Mars mineral-mapping from orbit has detected minerals that can indicate where a volcano erupted beneath an ice sheet. The site is far from any ice sheet on modern Mars, in an area where unusual shapes have been interpreted as a possible result of volcanism under ice. Credit: NASA/JPL-Caltech/JHUAPL/ASU

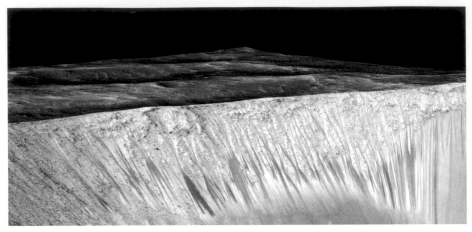

HiRISE has been monitoring the recurring slope lineae (RSL) over the mid-latitude and equatorial regions of Mars. One of these sites is a crater on the floor of Melas Chasma. Credit: NASA/JPL/University of Arizona

Data from MRO have also provided information about Mars' past, showing three distinct periods on Mars. Observations of the oldest surfaces on the planet show that diverse types of watery environments existed—some more favorable for life than others. More recently, water cycled as a gas between polar ice deposits and lower-latitude deposits of ice and snow, generating patterns of layering linked to cyclical changes similar to ice ages on Earth.

Scientists used the orbiter's mineral-mapping spectrometer CRISM to discover that volcanoes erupted beneath an ice sheet on Mars' southern hemisphere billions of years ago, far from any ice sheet on the Red Planet today. The research shows there was extensive ice on ancient Mars and it also adds information about an environment combining heat and moisture, which could have provided favorable conditions for microbial life.

Perhaps the most intriguing of MRO's discoveries shows the possibility of liquid water being present on Mars today. Using the imaging spectrometer and HiRISE, researchers saw mysterious dark streaks on slopes that appeared to ebb and flow over time. They would darken and appear to flow down steep slopes during warm seasons, and then fade in cooler seasons.

These downhill flows are called **recurring slope lineae** (RSL) and provide evidence of salty, briny liquid water flowing down the slopes. Scientists say the salts would lower the freezing point of a liquid brine just as salt on roads here on Earth causes ice and snow to melt more rapidly. This briny water is likely a shallow subsurface flow, with enough water wicking to the surface to explain the darkening.

"These RSLs are the biggest surprises scientifically of the mission," said Zurek, "and they help us appreciate how much Mars—a planet that has changed greatly over time—continues to change today."

OF TWO MINDS

Virtually all spacecraft incorporate redundant systems, meaning they have two identical systems on board, with one used as the primary system and another as a backup. Once a spacecraft is out in space—with the exception of the Hubble Space Telescope—there's no way to go out and fix the hardware, so having a backup on board is like an insurance policy.

Engineers in the Mars Reconnaissance Orbiter mission control room. Credit: NASA/JPL-Caltech

There can be redundancy in the computer, electronics and other key components. The computer for MRO, the command and data-handling subsystem, is essentially the "brains" of the orbiter and controls all spacecraft functions. The main computer is known as the A-side computer, and the backup is the B-side.

"There is kind of an unwritten rule that you don't swap sides unless you really have to," Zurek said, "because it might have been years since you last used the B-side, and you don't actually know if it is working."

MRO has a solution for that. It readily switches back and forth between side A and side B all on its own. Without warning. Frequently.

"We like to say that MRO has its own mind or a split personality," Johnston said, "and we don't really know why. We've investigated it and have never really been able to really determine why it jumps back and forth between the two main computers, and to be honest, we have no control over it."

Despite MRO's split personality, the spacecraft continues to function well, Johnston said. "Luckily both personalities are good ones, for the most part."

However, whenever this hiccup happens, the spacecraft goes into safe mode, so all activities come to a halt. The spacecraft then orients itself so the solar panels point toward the Sun (first order of business is to have power), and then it points the high-gain antenna toward Earth so it can receive commands.

Then the team can perform diagnostics, possibly figure out what went wrong or prompted the side swap, and bring MRO out of safe mode. This can take several days, however, which means the science instruments can't do their jobs, and MRO can't act as a relay for the rovers. So, MRO's flip-flopping affects the other missions, too. Most of the time the Odyssey orbiter can step in to provide the relay, but its data rate isn't as high.

While most of the side swaps have been mostly an inconvenience, there was one point in the mission when it started happening with alarming frequency.

"We had four events from 2007 to 2008, and then in 2009 we had four events that took place in close succession," Johnston said. "It looked like it was going down a path where these events were happening faster and faster and that we may not be able to control the spacecraft if this continued."

They took a long time to study the problem, standing down from science and other aspects of the mission. While they didn't completely solve the side-swap issue, they identified a related problem that potentially could have been fatal.

"We found it was possible that under the right set of circumstances the spacecraft could lose its memory, completely," Johnston said, "in which case it would revert and think it was back on the launchpad, and it would only want to communicate [or get power] using the hardwired umbilical that is used when the spacecraft is loaded on the rocket on the pad. So, we had to go in and do a special software change that tells the spacecraft it is now only in its science phase for ever and ever. No going back to the launchpad phase."

MRO has been doing better since that fix, and Johnston said the team is now "more confident that when these events happen we aren't going to lose the spacecraft, but just have the 'personality changes' from side to side."

But things do happen on an aging spacecraft. The day I visited Johnston and Zurek at JPL, the radio on board MRO that performs the data relay for the landers quit working. The team re-cycled the radio (turned it off and on) and it regained operation.

"So, even as we speak, even now after ten years of operations in deep space, there are surprising events," Zurek said. "Things do age, space is a pretty harsh environment by and large, and I think this orbiter going around Mars almost every two hours, thirteen times a day, making over 45,000 orbits now, it's quite incredible that MRO keeps righting itself."

COMET FLYBY

In early 2013, astronomers discovered a new comet named C/2013 A1 Siding Spring. Shortly after its discovery, scientists realized it was heading toward Mars. After the likelihood of the comet actually hitting Mars was ruled out, NASA decided to make extensive plans to study the comet with its spacecraft fleet at Mars, which included MRO. Johnston said this was both exciting and nerve-wracking, as something like this had never been done before.

"Anytime you're doing things that are new and different, you do get nervous about those kinds of events," he said. "When Comet Siding Spring flew by Mars in October of 2014, that was a very unusual event. For an orbiter that is designed to look at the surface of the planet, we decided to let HiRISE do a lot of imagining of the comet as it went by. To me, that is one of the cooler aspects of the mission, where when something unexpected happens, we're able to take advantage of it."

This artist's concept shows NASA's Mars orbiters lining up behind the Red Planet for their "duck and cover" maneuver to shield them from comet dust from the close flyby of comet Siding Spring (C/2013 A1) on Oct. 19, 2014. The comet's nucleus missed Mars by about 87,000 miles (139,500 km), shedding material as it hurtled by at about 126,000 mph (56 km/s), relative to Mars and Mars-orbiting spacecraft. The NASA orbiters at Mars are Mars Reconnaissance Orbiter, Mars Odyssey and MAVEN. Credit: NASA/JPL-Caltech

Comet Siding Spring Plans for Mars Orbiters:
Duck and Cover!

To photograph a fast-moving target from orbit, engineers for MRO at Lockheed Martin in Denver precisely pointed and slewed the spacecraft based on comet position calculations by engineers at JPL. To make sure they knew exactly where the comet was, the HiRISE team photographed the comet twelve days in advance, when it was barely bright enough to register above the detector's noise level. To their surprise, it was not exactly where orbital calculations had predicted it to be. Using the new positions, MRO succeeded in locking onto the comet during the flyby. Without this double check, its cameras may have missed seeing Siding Spring altogether.

Once the comet safely flew past, there were still concerns, however.

"The concern about that was even though it's great to do the observations and see the comet debris trail," Johnston said, "we were really concerned that the particles in the debris might provide micrometeorite strikes on the spacecraft, which could even kill the spacecraft. We synchronized our orbit so that as the comet flew by and the particle stream was behind it, Mars would go through that and we would be right behind Mars, and therefore we wouldn't be exposed to it head-on."

Even though it all worked out, Johnston said he was extremely nervous during the event. "Even if you've done the analysis and done everything you can to protect the spacecraft, it might come down to what the spacecraft gods might say for you that day," he said with a smile.

YOU CAN TAKE A PICTURE ON MARS

In the decade that it's been snapping photos, HiRISE has captured close to 50,000 images of Mars' surface. In doing so, it has provided a tremendous service to researchers throughout the scientific community. The only downside (if you can call it that) to HiRISE is that its images are so close-up, they cover only very small areas. That means no matter how long the spacecraft "lives," there's no way it could photograph all of Mars. In fact, after ten years, despite all the images it has taken, HiRISE has imaged just 2 percent of Mars' surface.

From the beginning, McEwen wanted to make HiRISE "the people's camera," so there are several ways for the public to participate in the mission. One allows you to suggest imaging targets for HiRISE.

"We have to choose our targets carefully," said Ari Espinoza, who leads the HiRISE outreach and media projects. "So, the HiRISE team invites the public to participate in the choices. The program is called HiWish, and it's accessible to professionals and amateurs, grownups and kids, and anyone around the world."

You just need to log in on the HiWish Web site and suggest a target. If you don't know where to suggest, there is helpful information and a map you can zoom in and out to help you choose. The science team will let you know if your target was chosen.

Another project is called BeautifulMars.

"Since we've always considered HiRISE the people's camera, I wanted to extend that concept to people around the world, even if they don't speak English," said Espinoza. "The HiRISE science team has written over 1,700 captioned images, basically written for the public, and we started enlisting as many volunteers as we could find who were interested in learning about Mars and wanted to translate captions. Now it has just taken off."

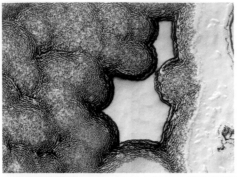

The South Polar residual cap (the part that lasts through the summer) is composed of carbon dioxide ice. Although the cap survives each warm summer season, it is constantly changing its shape due to sublimation of carbon dioxide from steep slopes and deposition onto flat areas. This observation was acquired on March 23, 2015. Image Credit: NASA/JPL-Caltech/Univ. of Arizona

What appear to be trees growing from the Martian surface are actually dark streaks of collapsed material running down sand dunes due to carbon dioxide frost evaporation. Credit: NASA/JPL/University of Arizona

Espinoza said they are very proud of the BeautifulMars Project, and HiRISE is now represented in 27 languages, the most of any active NASA mission. "It's been successful to the point where our materials can help in language revitalization efforts," he said, "which has caught the attention of language councils in Northern Ireland and Cornwall, for example. We certainly hope to add more, including First Nations languages."

The team also uses the power of social media to share their images in as many languages as possible. "We like to think of ourselves as masters of social media, using as many avenues as possible to get HiRISE and our images to the widest possible audience," Espinoza said.

They also have an ebook series, a new outreach initiative called MarsPoetica, where people can submit poems about Mars, and they have a YouTube channel called HiClips that shares several images in a video format.

Since the HiRISE images are so large and can include long strips of imagery, it can be difficult to look at the images. The team created a special web viewer, called HiView, to easily view the large images.

"HiView lets you navigate these enormous, incredibly high-resolution images of the Martian surface very quickly," Espinoza said. "You can zoom into an area or zoom out to see the larger view. It's designed as a way to deal with these very large images for viewing them efficiently."

This was originally developed for the scientists to be able to navigate through the images, but it's great for anyone who is interested in exploring Mars close-up.

There's also a citizen science project from the Cosmoquest collaboration (www.cosmoquest.org) called Mars Mappers that uses HiRISE data to look for craters. Here, people can help scientists create a global impact crater database of Mars. This helps them understand not only the rate of impacts on Mars surface, but can also assist in determining the ages of the different surface regions on Mars.

MRO'S LEGACY

While HiRISE imagery provides detailed close-ups of Mars, another camera, the CTX, provides a six-meter-per-pixel resolution, better than any other previous camera. CTX has imaged nearly 97 percent of Mars' surface, allowing the science team to create quite detailed maps of Mars. Other instruments continue to study Mars' surface, interior and atmosphere.

How long can MRO continue?

"We have no end of mission in sight," said Johnston, "and NASA wants us to keep flying. Our primary goal is to fly and be there for the Mars 2020 rover mission and to be their relay satellite." (Read about that mission on page 212.)

Johnston said MRO has enough fuel for its orbital maneuvers to last at least another twenty years.

"That was one of the development options because we had a large launch vehicle," he said, "so we were literally able to top off the fuel tanks at the Cape when the spacecraft was being processed for launch. That has given us a tremendous resource, and fuel is a commodity that we are rich with. I think the things we worry about more are mechanical cycles, like the gyros on the reaction wheels are moving all the time, and things like that will soon start to approach their designed life cycles. But for the most part, the spacecraft is really healthy and doing very well."

Johnston said he clearly believes that MRO is the flagship of the fleet of spaceships at Mars. "Everything we do helps the Mars program in context and allows our scientists to excel and be the world-class scientists they are."

While the path of getting MRO to Mars wasn't always easy, Zurek said they are now reaping the benefits of a long-term mission.

"There is value in long-term records," he said, "however, we aren't in this business of exploration to do the same thing again and again. We always want to learn more."

Zurek said the longevity of the spacecraft is a testament to the teams that built and operate MRO. At one point in our conversation, he realized he was speaking about the spacecraft as if it were a person. He laughed and said, "So, I even talk like it has a mind and listens to us. In a sense it does. MRO has been a good friend to us for a long time."

CHAPTER 9

SHOOTING THE MOON: THE LUNAR RECONNAISSANCE ORBITER

EVIDENCE OF APOLLO

In July 2009, it was just days before the 40th anniversary of the first landing of humans on the Moon, the Apollo 11 mission on July 20, 1969, when Neil Armstrong and Buzz Aldrin (with crewmate Mike Collins orbiting above) took their "giant leap" for all mankind.

Mark Robinson had an idea for a very special way to celebrate this historic anniversary, but he didn't know if it was going to work. However, he and his team were the only ones in the world—or in the solar system—that could make it happen.

Just four weeks earlier, the Lunar Reconnaissance Orbiter (LRO) launched from Earth and went into orbit around the Moon to map the lunar surface and explore our nearest neighbor in exquisite detail. On board the spacecraft was an imaging system called the Lunar Reconnaissance Orbiter Camera (LROC), which includes three cameras that capture high resolution black and white images and moderate resolution multi-spectral images of the lunar surface.

Sunrise on the Moon creates long shadows from the "central peak" mountain in Tycho crater. The entire mountain complex, shown in this image from the Lunar Reconnaissance Orbiter, is about 9.3 miles (15 km) wide, left to right. The peak reaches a maximum height of about 1.2 miles (2 km) above the crater floor. Credit: NASA Goddard/ Arizona State University

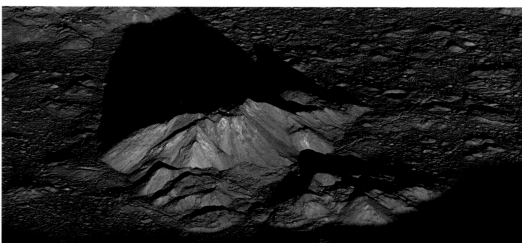

Artist's concept of Lunar Reconnaissance Orbiter. Credit: NASA

Robinson is the principal investigator for LROC, leading the team of imaging specialists on the mission. The plan for the Apollo 11 anniversary was to try and take images of the Apollo landing sites with this new, high-resolution camera. Previous orbiters had tried to locate and image the relatively small artifacts left on the Moon by the Apollo astronauts—the Apollo lander descent stage, various experiments and equipment and the lunar rovers for the later Apollo missions—all of which remained on the Moon after the astronauts returned home. But no orbiter had been successful in capturing evidence of the first and only series of human visits to a planetary body other than Earth.

The other challenge for Robinson and his team was that LRO was just settling into its orbit. When it first arrived at the Moon in late June 2009, it went into a highly elliptical orbit that would be refined into a more circular and closer lunar orbit. The spacecraft and instruments were being checked out in this initial orbit, and it would be several weeks until LRO reached its final mapping orbit. That meant the spacecraft and the cameras might not be as close as Robinson thought was needed to see the Apollo landing sites this early in the mission.

Also, the team was still commissioning the cameras; testing them, making sure all the equipment and systems were working correctly. But time was short if the images were going to be taken, sent back to Earth and processed in time for the July 20 anniversary. So Robinson and his team sent commands for LROC to attempt images when the spacecraft flew over the right locations.

"The LROC team anxiously awaited each image," Robinson said back in 2009. "We were very interested in getting our first peek at the lunar module descent stages just for the thrill – and to see how well the cameras had come into focus."

What they saw was indeed a thrill.

"The images were fantastic and so was the focus of the cameras," Robinson said. "It was great to see the hardware on the surface, waiting for us to return."

The pictures show the Apollo missions' lunar module descent stages for Apollo 11, 15, 16 and 17 sitting on the Moon's surface. The resolution wasn't completely crisp enough to show details of the descent stage, but long shadows from a low Sun angle made the modules' locations clearly evident. But amazingly, also visible at the Apollo 14 landing site were the tracks left by the astronauts as they walked repeatedly in a "high traffic zone" and perhaps by the Modular Equipment Transporter (MET) wheelbarrow-like carrier used on Apollo 14, a mission that didn't have a lunar rover.

"These first images were taken when LRO's orbital altitude was about two times higher than it would be during the normal 50-kilometer (31-mile) mapping orbit," Robinson said from his office at Arizona State University (ASU) in 2016. "From that altitude we knew would see the descent stages and possibly the ALSEP packages [ALSEP, the Apollo Lunar Surface Experiments Package, a collection of experiments designed to monitor the environment and make scientific measurements at each Apollo landing site]. The Sun was very low to the horizon resulting in long dramatic shadows being cast by the hardware. You could see that the Apollo 16 lunar module had landed right on the edge of a small crater, and the LM shadow reached across the crater shadow up onto the illuminated far wall."

It was exciting for Robinson and his team—and all of NASA—to see the landing sites because it showed how well the camera was operating. "We were very excited about the quality of the data," Robinson said.

LRO's first views of lunar modules at Apollo 11, 15, 16 and 17 landing sites. Credit: NASA/Goddard Space Flight Center/Arizona State University

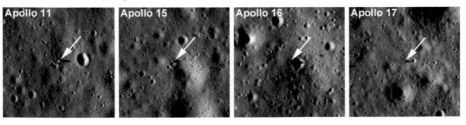

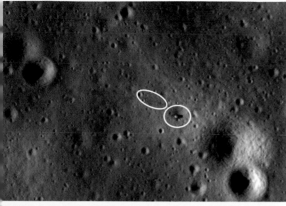

The first image of the Apollo 14 landing site had a particularly desirable lighting condition that allowed visibility of additional details. The Apollo Lunar Surface Experiment Package, a set of scientific instruments placed by the astronauts at the landing site, is discernable, as are the faint trails between the module and instrument package left by the astronauts' footprints. Credit: NASA/Goddard Space Flight Center/Arizona State University

LRO's first view of the Apollo 11 landing site, the first human mission to land on the Moon. The shadow in the center of the image, circled, reveals the location of the lunar module descent stage. Credit: NASA/Goddard Space Flight Center/Arizona State University

The rest of the LRO team was pretty stoked, too.

"The cameras are operated out of ASU," said Rich Vondrak, who was the LRO project scientist from 2007 until his recent semi-retirement, "and instrument teams are located around the US, but the mission operations team is here at the Goddard Space Flight Center in Maryland. So, we knew Mark and his team were trying to take these images, and I got a message from him late one night. I ran down and looked at my computer and when I saw the images, I said, 'Holy cow! You can actually see the hardware! I called my wife Mary over and it was just breathtaking. Not only do we see the hardware, we can also see where they walked."

Vondrak noted how the individual footprints can't be seen because they are smaller than the camera's resolution. "But we know from the video and pictures taken by the astronauts that when they walked on the Moon, because of the lower gravity and their big bulky spacesuits, it actually worked better for them to bounce across the surface and that kicked up a lot of the surface material. It would be like when you walk across the lawn with an inch (25 mm) of snow, and since you kick up the snow, you leave a track that is bigger than individual footprints."

For Vondrak, who had been part of the Apollo science team that assisted the astronauts on their missions, seeing the landing sites was especially thrilling.

"Being able to take these observations and relate them to important events in human history I found to be exciting, especially when I was one of the first people to see these images," he said. "It was very exciting and fun."

For Noah Petro, currently the LRO deputy project scientist, seeing these first images from the Apollo sites was very personal.

"My father was an engineer for Apollo and worked on parts for the lunar module and the backpacks, the portable life support system," he said. "The backpacks were all left on the surface of the Moon because they were extra weight NASA didn't want to haul back to Earth. I remember looking at those first images and trying to determine which pixels had the backpacks. That's a great memory I'll always have about the first time I saw those images from LROC. Plus, I've always been an Apollo fan, even as a kid I was in awe of what Apollo accomplished."

Across the world, those images created a buzz of excitement, both for the generations of people who recalled those poignant moments in history when humans went on such a grand adventure, and for the generations who were too young to remember those days of Apollo. And then there was that small but persistent group of conspiracy theorists who believe—despite the mountains of evidence to the contrary—NASA never landed on the Moon, that the space agency staged the whole thing in a warehouse somewhere. But of course, as many have pointed out, it would actually be harder to fake the Moon landings and keep the evidence secret all these years than to have actually landed on the Moon. Furthermore, if you were going to fake a Moon landing why would you do so seven times, including one attempt (Apollo 13) that was a near disaster?

"GOING BACK TO THE MOON TO STAY"

LRO's journey to the Moon began in January of 2004. President George W. Bush made a major policy speech, saying that NASA was going to return astronauts to the Moon in preparation for going on to Mars. But the first step would be a lunar orbiter to survey the Moon in great detail for the future landers and human missions.

Artist's conception of the Orion spacecraft from NASA's Constellation Program in lunar orbit. Credit: NASA

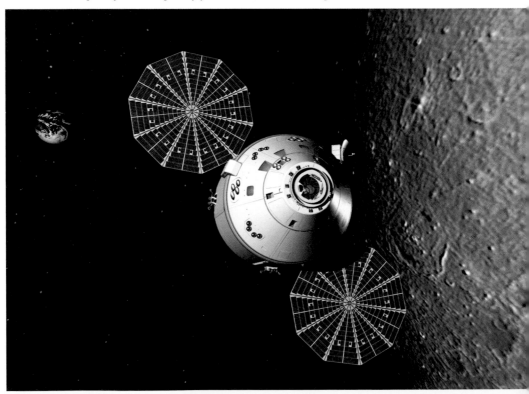

"Our goal," Bush said at a press conference at NASA, "is to return to the moon by 2020, as the launching point for missions beyond." He proposed sending several robotic probes to the Moon by 2008, with a human mission as early as 2015, "with the goal of living and working there for increasingly extended periods of time." A new program called Constellation would build a big new rocket to launch humans to the Moon and destinations beyond.

"LRO was established as the very first mission," said Vondrak, "and we had just four years to put the mission together. The plan was to have an impressive set of missions, with an orbiter and a lander and presumably more to follow and then soon thereafter have astronauts back on the surface. The robotic missions would map the Moon to find the best landing sites, search for resources and make other measurements that would enable safe visits by astronauts."

When the LRO mission was conceived it was the highest priority. "We had a presidential mandate, 'thou shalt go to the Moon with an orbiter, followed by a lander, and build up a program to send humans back to the Moon,'" Vondrak said dramatically. "And much of that should have happened by now, as humans would be launched on the Orion spacecraft, with the goal of landing by about 2020. NASA headquarters actually wanted to do it before the 50th anniversary of Apollo 11."

Back then, with the excitement of the mandate, Vondrak said it was a stressful but exhilarating atmosphere under which to work.

"Usually things are very bureaucratic, with it taking several years just to get a mission approved and started," he said. "Instead we had just a few months to get everything organized, to write up our mission plans, get it released and get all the signatures required."

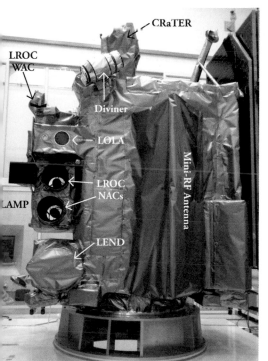

But Vondrak and the team at NASA Headquarters quickly set up the program, issued the announcements for instruments, and got the program running in about six months.

"Just the year before, the MRO mission had been vetted and approved," Vondrak recalled, "and since our mission was going to be similar—to map out the surface in high resolution and do other investigations of the surface—we just modeled our LRO proposal plan after MRO."

But Vondrak also worked with a great team of scientists and engineers.

LRO in final stages of preparation before launch in May 2009. Credit: LROC/Arizona State University

"We had an excellent team here at Goddard," he said, "Craig Tooley is the best program manager I've worked with, and he put together a fine team that built the spacecraft at Goddard. Like any high priority, urgent project, you have a committed team, everyone worked hard, and team morale was high. If there were problems, everyone was very dedicated and very creative. That's the way to get things done. Take a high performance team, give them the tools they need, give them a sense of urgency, and it's remarkable what you can accomplish."

Because of the high priority, there was no shortage of people wanting to get involved, so the science team was easy to assemble. "The scientists knew it was not only important for the country, but it would be one of the finest missions to the Moon," Vondrak said, "and there were many people in the science community who were waiting for this type of mission."

The team was assembled, the spacecraft was built and LRO launched on June 18, 2009, NASA's first unmanned Moon shot since 1998, returning to the Moon just under a month before the 40th anniversary of the Apollo 11 lunar landing.

About the size of a Mini Cooper car, the $504 million LRO began its mission with the goal of spending at least one year mapping the Moon for future manned missions, as well as spending several more years conducting science surveys.

"I remember when LRO arrived at the Moon, I was quoted in a press release saying, 'We are in lunar orbit, and we're going there to stay,'" Vondrak said.

LRO has performed perfectly and much longer than its one-year prime mission. But less than a year after LRO's launch, NASA's plans changed.

WHERE ARE WE GOING?

In February 2010—after the United States had sustained a stock market crash and lending crisis, and with the country facing a federal deficit of $1.26 trillion—President Obama announced that the planned Constellation Program would be cut, and NASA would not be sending humans to the Moon, at least not any time soon. Since then, the future of NASA's human spaceflight program has been in limbo.

"What happened was, the lack of funding got in the way," Vondrak said, "so it was clear that the costs of going to the Moon and then going on to Mars were difficult to identify and sustain. So not only was Constellation cancelled, the planned robotic lunar exploration program was cancelled after LRO."

And still to this day, NASA doesn't have a clear vision of where its human spaceflight program will go beyond low Earth orbit. While ideas for missions to an asteroid have been tossed around, Mars has been the ultimate destination in every space enthusiast's mind, and NASA now says

Mars is the definitive goal. However, landing on the red planet with human-sized spacecraft and payloads is extremely expensive and difficult, with many technologies for such a mission having yet to be developed, much less understood. All the equipment and procedures needed for a human presence on the distant Mars will require a nearby 'proving ground' of sorts to test them out. The Moon offers that proving ground, so at some point a return to the Moon seems inevitable on the road to Mars.

Plus, the funding for any human mission to Mars is elusive at best. The amount of funding provided to NASA doesn't allow for big programs—like the Apollo program in the 1960's—to take place today. And unlike Apollo, there is no sustained presidential mandate to get to Mars "in this decade." The potential destinations for NASA's human spaceflight program seem to change with every presidential administration.

But while NASA—and the politicians who help decide on the agency's funding—determine where human space missions might go, LRO still circles the Moon, taking data and making discoveries that are changing our view of the Moon. We now know the Moon is not the dry, dead world it was once thought to be.

A visualization of the LRO spacecraft as it passes low over the Moon's surface near the lunar South Pole. Credit: NASA/GSFC/SVS

"Both internationally and here in the U.S., scientific interest in the Moon remains high and we're still making discoveries," said Petro. "I just got an email last night about a really interesting finding, and something like that happens every week. We're certainly not running out of science! And we have this great spacecraft that we can interact with, so if you need more data, you can talk to Mark and his LROC team to get more images, or the other instrument teams to get more observations. That's a great asset."

With its seven different science instruments, LRO has produced a tremendous data volume, the largest data set of any planetary science mission (which doesn't include SDO) and more data than all other planetary missions combined.

"That is very remarkable and it is all freely, publically available," Petro said, "so, the data is there for the world to use, and we encourage people to do that."

LRO has been operating at the Moon for more than seven years, and when I talked with Vondrak and Petro they were awaiting a decision by NASA for a mission extension for two more years. On July 1, 2016, they recieved word the extension had been granted.

"Even though it was designed for a relatively short life, the spacecraft was built with care by an excellent team," said Vondrak. "We have not had any significant problems; all our systems are working fine. Plus we have a wonderful science team that has made some exciting discoveries."

By careful management of the spacecraft and its resources, the LRO team has been able to maintain all their consumables such as fuel and the mission could keep operating for another six to eight years, if NASA would approve that long of a mission, Vondrak said.

"Hopefully people go back to the Moon someday," he said, looking optimistically to the future, "and LRO will help them when they return. We have a legacy where we are compiling the basic 'guidebook' to the Moon, a handbook, that I'm sure 40–50 years from now when we have all sorts of people from many nations exploring the Moon, they will look to the database and the maps we've created to guide them on their way."

Vondrak said because of LRO, we now know the Moon is a very interesting place.

"Before, it was just that faraway place," he said. "We had the Apollo pictures of the equatorial regions, and images from other orbiters, but they just can't match the resolution of our camera and the superb quality of our other measurements. Now, if you want to go anywhere on the Moon, we'll tell you all about it. That's the exciting thing about planetary science: it lets us know that these aren't just objects out there in space, they are special places."

MOON BEAMS

LRO's laser mapping instrument zaps the Moon an incredible 140 times every second, measuring the ups and downs, nooks and crannies on the lunar surface to an accuracy within four inches (10 cm).

"We can provide topographic maps of the Moon that have finer grid spacing than the hiking maps at the US National Parks," Vondrak said. "We know what the Moon looks like in tremendous detail, and we actually have better knowledge of the shape, contours and topography of the Moon than any other object in the Solar System. That includes the Earth, because most of the Earth's surface lies beneath the ocean, and the seafloor is not mapped as well as the Moon."

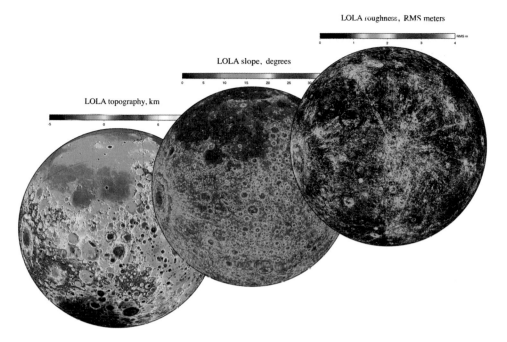

LOLA data provides complementary views of the near side of the Moon: the topography (left) along with new maps of the surface slope values (middle) and the roughness of the topography (right). Credit: NASA/LRO/LOLA Science Team

This mapping system, called the Lunar Orbiter Laser Altimeter (LOLA) has ten times greater vertical accuracy and 300 times more measurements than any previous laser altimeter mission. In addition to creating detailed contour maps, an interesting detail about how this instrument works provides further evidence of Apollo's equipment left on the Moon.

The Laser Ranging Retroreflector (LRR) experiment was deployed by the astronauts at the Apollo 11, 14 and 15 sites. It consists of a series of corner-cube reflectors, which reflect an incoming light beam back in the direction from which it came. A similar device was also included on two of the Soviet Union's Lunokhod rovers, which landed on the Moon in 1970 and 1973. For the majority of the mission, LOLA needed to be turned off when it passed over the Apollo and Lunokhod sites because bouncing the laser off any of the retro-reflective mirrors could have damaged LOLA.

However, in 2017 LRO's orbital altitude will be increased and the plan is to use LOLA at the Apollo and Lunokhod landing sites so that topographic information can be obtained for these areas and be incorporated into the detailed lunar contour maps. Petro said LRO's increased altitude along with the LOLA laser decreasing in strength over time, decreases the chance of any damage to the instrument.

Ever since the LRR experiments were deployed, the McDonald Observatory in Texas has beamed a laser at these mirrors and measured the round-trip of the beam. This provides accurate data on the Moon's orbit, the rate at which the Moon is receding from Earth (currently 3.8 centimeters [38 mm] per year) and variations in the rotation of the Moon. These are the only Apollo experiments that are still returning data.

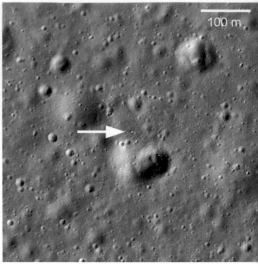

The Laser Ranging Retroreflector experiment deployed on Apollo 11. Credit: Lunar and Planetary Institute

Soviet robotic lander Luna 17 still sitting on Mare Imbrium where it delivered the Lunokhod 1 Rover in November 1970, imaged by LROC. Credit: NASA/GSFC/Arizona State University

And so while some people debate the reality of the Apollo landings, scientists from LRO must deal with the practical consequences of the items left on the Moon by the Apollo astronauts and other lunar explorers.

THE NEW MOON

2009 ended up being the year that marked a huge change in our understanding of the Moon, thanks to several different missions. Prior to that year, the Moon was believed to be extremely dry, primarily because of the lunar samples returned during the Apollo program. While many Apollo samples contained some trace water or minor hydrous minerals, these findings were typically attributed to terrestrial contamination since most of the boxes used to bring the Moon rocks to Earth were not air-tight. This led the scientists to assume that the trace amounts of water they found came from Earth air that had entered the containers. From then on, the assumption remained that, outside of possible ice at the Moon's poles, there was no water on the Moon.

Forty years later, an instrument on board India's Chandrayaan-1 spacecraft, the Moon Mineralogy Mapper (M^3) found that water exists diffusely across the Moon as water molecules or hydroxyl—or both—adhering to the surface in low concentrations. This was confirmed in data from two other spacecraft flying by the Moon; the repurposed Deep Impact probe on its way to rendezvous with a comet, and Cassini spacecraft during its Earth flyby in 1999. Roger Clark from the M^3 team reanalyzed archival data from observations Cassini made of the Moon, and that data as well agreed with the finding that small amounts of water appear to be widespread across the lunar surface, likely brought there by the solar wind.

"Then the samples from Apollo were re-analyzed and found to have small amounts of water inside," said Petro, who was part of the M^3 team. "And so LRO launched at this sort of turning point at our understanding of volatiles at the Moon and has really helped progress that story. Plus, there was LCROSS."

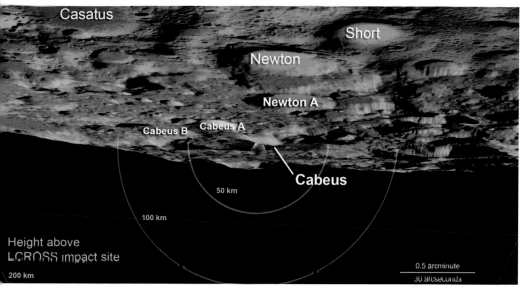

An annotated image of the Moon's south pole with the locations of several craters, including Cabeus, where LCROSS impacted. Credit: NASA/Goddard Space Flight Center Scientific Visualization Studio

WATER AND MUCH MORE

LCROSS, the Lunar CRater Observing and Sensing Satellite was a companion mission to LRO, with the mission objective of determining whether water ice could be found in what is called a **permanently shadowed region** (PSR), found in dark craters at the Moon's poles. The interiors of these craters rarely or never receive any sunlight because the craters' towering rims and the Sun's oblique angle casts long, dark shadows over much of the crater's interior.

LCROSS launched along with LRO, and flew to the Moon with the Atlas V's Centaur upper stage. The plan was for the spent rocket stage to impact into Cabeus Crater, a PSR at the lunar south pole. Four minutes behind, the "shepherding" LCROSS spacecraft would follow close on the impactor's heels, monitoring the resulting ejecta cloud with nine science instruments to see what materials were excavated from inside this dark, unstudied crater. LRO, the Hubble Space Telescope and telescopes back on Earth would also attempt to image the action.

The mission caused some apprehension by a few concerned citizens who were worried LCROSS would hurt the Moon, or that NASA was "bombing" the Moon. Even though the impact was expected to kick up tons of lunar regolith, LCROSS principal investigator Tony Colaprete estimated the impact would have about 1 million times less influence on the Moon than a passenger's eyelash falling to the floor of a 747 jet during flight.

"What we're doing with the Moon is something that occurs naturally four times a month on the Moon, whether we're there or not," Colaprete said. "The difference with LCROSS is that it is specifically targeted at a certain spot, Cabeus crater, and that the laws of physics mean there will be a miniscule perturbation."

On October 9, 2009, the mission went as intended, and although the plume wasn't as visible from Earth as was hoped, LCROSS flew through the debris plume, collecting and relaying data back to Earth before impacting the lunar surface itself and creating a second debris plume.

The 5,200 pound (2,400 kg) Centaur rocket created a crater about 25 to 30 meters (82 to 98 ft) wide, and the LCROSS team estimates that somewhere between 8,800 to 13,000 pounds (4,000 to 6,000 kg) of debris was excavated out of the dark crater and into the sunlit LCROSS field of view.

The LCROSS team was able to measure a substantial amount of water and found it in several forms. "We measured it in water vapor," said Colaprete, at NASA Ames, "and much more importantly, we measured it in water ice. Ice is really important because it talks about certain levels of concentration."

How much ice? Colaprete described it as "blocks" of ice. With a combination of near-infrared, ultraviolet and visible spectrometers onboard the shepherding spacecraft, LCROSS found about 155 kilograms (342 pounds) of water vapor and water ice were blown out of the crater. From that, Colaprete and his team estimate that approximately five to eight percent of the total mass inside Cabeus crater could be attributed to water ice alone.

Jennifer Heldmann, another scientist with LCROSS, said the instruments on their spacecraft along with LRO's observations—and in particular the LAMP instrument (Lyman Alpha Mapping Project)—the most abundant volatile in terms of total mass was water, followed by hydrogen sulfide, ammonia, sulfer dioxide, acetylene, carbon dioxide and several hydrocarbons.

"We found all these volatiles," she said, "which just means gasses that can condense out at very low temperatures. Some of our science team members think of these PSRs as the 'junk yards' of the Solar System, because material gets deposited there from impacts and other processes, and it's just so cold that those molecules don't have enough energy to get out again. So, there's a vast reservoir of water and all this other stuff that is stuck at the poles."

If humans ever return to the Moon, access to water and other elements would be an important resource. Lunar water could be used for drinking, or its components—hydrogen and oxygen— could be used to manufacture things like rocket fuel and breathable air.

Heldmann and Colaprete said LCROSS's findings have completely changed our understanding about the Moon's poles.

"It really was exploration," Colaprete said. "We were going somewhere we had absolutely never gone before, and scientists have wanted to study a PSR for decades. We weren't disappointed, although some of the findings still have us scratching our heads."

Other recent discoveries have shown additional surprises, as well as more resources on the Moon. In 2015, LRO's Lunar Exploration Neutron Detector (LEND) instrument discovered that hydrogen-bearing molecules—including possibly water—are more abundant on crater slopes in the southern hemisphere that face the lunar South Pole.

LAMP is an imaging spectrograph that maps the lunar surface and looks at the tenuous atmosphere of the Moon in ultraviolet wavelengths. It, too, detected water ice at the poles, and found helium in the lunar atmosphere, about 40,000 helium atoms per cubic centimeter. But that amount seems to vary cyclically with surface temperatures and the lunar day/night cycle. The team is still determining whether the helium comes from sources on the Moon or if it is brought to the Moon by the solar wind.

Additionally, LEND, LAMP and LOLA have detected changes to the surface of the Moon that may be related to what is called **volatile migration**, meaning the small amounts of water and other substances on the Moon's surface come and go and move around. While the mechanism for this is not yet well understood, it is thought to be related to, again, changes in temperature during a lunar day.

A radar instrument on LRO called the Miniature Radio Frequency (MiniRF) creates radar maps of the Moon, including the first radar maps of the lunar farside. Later in the mission, this instrument has had problems with its transmitter, so now the team is working in cooperation with the radio telescope at the Arecibo Observatory in Puerto Rico to make what are called **bistatic** measurements to look at the interior of the craters at the lunar poles. For these measurements, Arecibo transmits a radio signal that bounces off the Moon which is then received by Mini-RF.

"By making measurements at several angles between the two radar instruments, it is possible to get excellent data on subsurface ice," said Vondrak. "These are the first measurements like this attempted from Earth."

LRO has also found over 200 "caves"—steep-walled lunar pits—that are not only intriguing (future spelunking tours on the Moon?) but might offer shelter for future astronauts, protecting them from radiation, micrometeorites, and dust. They range in size from about 5 yards/meters across to more than 900 yards/meters in diameter. Exploring the caves could also provide more information about the interior of the Moon and how the Moon formed.

These images show various caves, called lunar pits, that LRO has found on the Moon. Each image shows an area about 728 feet (222 m) wide. Credit: NASA/GSFC/University of Arizona.

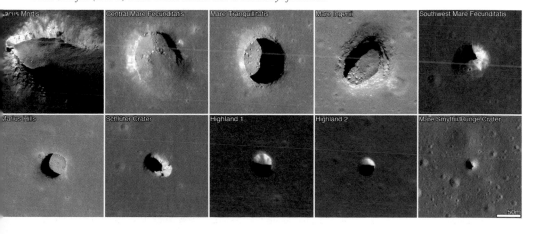

A new impact crater on the Moon found by LRO in coordination with researchers on Earth. Credit: NASA/ Arizona State University

A discovery by another of LRO's instruments, the Diviner thermal emission instrument, might not make the Moon seem very inviting for future explorers but it does give the Moon a certain distinction: it contains the coldest place in the solar system. The location for the shivering spot is—again—the permanently shadowed craters near the Moon's south pole. In their permanent darkness, they stay at a nearly constant -400 °F (-240 °C), or 33 degrees above absolute zero, colder than what's been measured at Pluto, and likely colder than the farthest reaches of our solar system.

Data from LRO has shown the Moon is actually shrinking. Newly discovered cliffs in the lunar crust called lobate scarps indicate the Moon shrank globally in the geologically recent past and might still be shrinking today.

"It's like if you take an orange or any fruit and set it out in the Sun, it starts to dry out, it develops wrinkles and cracks because the crust is bigger than the fruit as it is shrinking," said Vondrak. "That's what is happening to the Moon—although the process is extremely slow, but we see wrinkle ridges that are formed because of this slow shrinking."

One of the most amazing discoveries, Petro said, has been the camera team identifying new impact craters. "Impact craters are forming today, several within the last five to seven years," he said. "No previous mission has been able to do that, because the only way you monitor impact events is to be there for an extended period of time. No prior mission has been there for longer than about two years."

It turns out, lunar impacts happen more frequently than anyone expected. A research team at NASA's Marshall Space Flight Center in Huntsville, Alabama monitors the Moon for impact events, and in combining their data along with the LROC's, hundreds of impacts are being detected every year. The LROC team has been going back and looking at images taken in the first year or two of the mission and then comparing them to recent images to find new impact craters that were formed between the times the two images were acquired. Called **temporal pairs**, these before-and-after images enable the search for new impact events, as well as a range of other surface changes.

One impact event stands out as the brightest flash ever detected by the team at Marshall. On March 17, 2013, an object the size of a small boulder hit the surface in Mare Imbrium and exploded in a flash of light nearly 10 times brighter than anything ever recorded. The LROC team was able to obtain observations of the approximate location of the impact, and found the crater by comparing previous images.

Scientists estimate the meteoroid weighed 90 pounds (40 kg), measured about one foot (0.3 meters) wide and hit the Moon traveling 56,000 mph (90,000 kmh). The resulting explosion packed the equivalent of 5 tons of TNT. The crater itself is relatively small as lunar craters go, as it measures 61.7 feet (18.8 meters) in diameter. But the impact debris flew for hundreds of meters. LROC found more than 200 related surface changes up to 19 miles (30 km) away.

"People thought Moon was static—that it is no longer changing," said Vondrak, "but we are discovering there are changes. By re-imaging places, we have identified these fresh meteor impacts. We've found evidence of water ice in craters at the lunar poles, as well as polar hills that are almost constantly illuminated by the Sun, which would be a great place for future human explorers to live, because of the resources available there."

"Thoughts that the Moon is dead, well, it just isn't true," Petro said. "We do see changes on the Moon, and actually this is something I try to get people to think about, that in terms of the time LRO has been at the Moon, we've only been there for about 80 lunar days. While that's seven Earth years, that really isn't the way to keep track of it anymore, in my opinion. We see there are variations occurring on the Moon on a scale of a lunar day (29.5 Earth days). To really track and monitor the lunar processes, you need to be there for an extended period of time."

Even with the mission extension, LRO will be able to study the Moon for just 100 lunar days.

"There's more science to do, we're not anywhere near being 'done' with studying the Moon," Petro said. "By staying at the Moon for an extended period of time you're really able to capture those subtle variations and changes. These changes are really fundamental to any airless body, so not only the Moon but asteroids, the moons of Mars, and Mercury, for example. So I think our findings have not only changed the way people think about how these processes work on the Moon, but other airless bodies as well."

Spectacular oblique view of a 4,593 foot (1400 m) diameter crater that formed on the rim of Chaplygin crater. Delicate lacy fingers of ejecta highlight the hummocky and steep topography around this young crater. The very brightest material was excavated from the lower reaches of the crater and was some of the last material to exit the rapidly forming crater. Credit: NASA/GSFC/Arizona State University

THE MOON UP CLOSE

If you've ever looked up at the Moon on a clear night and wondered what it would be like see it up close, you're in luck. The incredibly beautiful and detailed images taken by LROC can almost make you feel like you're right there, in close orbit around the Moon.

"LROC takes all these awesome pictures of the Moon, and really, it makes the whole mission rewarding because of the Moon's breathtaking beauty," Vondrak said.

LROC consists of two narrow angle cameras (NACs) designed to provide 0.5 meter (1.6 ft)-scale black and white images over a 3 mile (5 km) swath of the Moon's surface, and one wide angle camera (WAC) that provides images at a scale of 100 meters (328 ft) per pixel in seven colors bands over a 37 mile (60 km) swath. The difference in the amount of surface area each camera covers is demonstrated by the fact that the WAC creates a new global map of the entire lunar surface every month, while the NACs—as of 2016—have taken images of about 40% of Moon.

The images from LROC provide a wealth of science data while capturing the stark beauty of the Moon.

"The LROC cameras were carefully designed and built by Malin Space Science Systems (MSSS) in San Diego," Robinson said. "Really, it's the dedication, care and attention to detail of the MSSS team that makes LROC such a superior imaging system."

Petro agrees, but adds that another aspect of what makes this suite of cameras so special is the people who operate the instrument.

"The team is great, and Mark Robinson is amazing," Petro said. "He has an eye for anticipating beautiful shots. For example, he knows if we slew the spacecraft to this point we can get a crater's central peak illuminated in a certain way. While he's thinking primarily about the science we can get out of the instrument, he's also been able to get these amazing pictures that are incredibly beautiful."

In this striking view of the Giordano Bruno crater, the height and sharpness of the rim are evident, as well as the crater floor's rolling hills and rugged nature. Credit: NASA/Goddard/Arizona State University

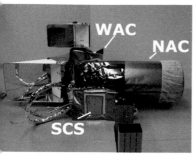

Annotated picture showing one of the NACs, the WAC and the SCS, a small electronics system that controls the three cameras. Credit: ASU/ LROC

Variations in Sun angles bring out details of the lunar landscape, with striking differences in bright and dark areas. Shadows play across boulder-filled regions, sunlit mountaintops stand out and against dimmer surroundings and craters yawn wide, inviting you to look inside at all the details LROC has captured. These images show the Moon to be a wonderful world in its own right, as well as a world in change. Buzz Aldrin's phrase of the Moon's "magnificent desolation" is aptly fitting.

Looking at the global WAC views at different illuminations can highlight even subtle features. Even though LRO is speeding around the Moon at about 3,500 miles per hour (1,600 m per second), the images are extraordinarily crisp. LROC has an exposure time of 0.3 milliseconds, three times shorter than the fastest off-the-shelf cameras.

Every day, LROC makes approximately 400 observations: the NACs make around 300 observations and the WAC makes about 100. This amounts to about 50 GB of data that is sent to Earth each day. An electronic system on board LRO called the Sequence and Compressor System (SCS) supports data acquisition for both cameras.

About 90 percent of the work required to target, downlink and process images is automated. Robinson said without automation, they wouldn't be able to keep up with the firehose of data. Even with the extensive use of automation, approximately 20 to 30 LROC team members are still needed to operate the instrument at ASU's dedicated science operations center.

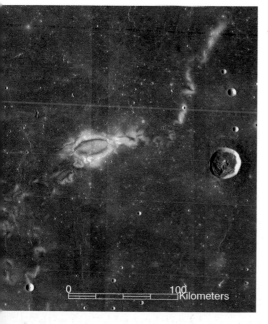

This is an image of the Reiner Gamma lunar swirl from NASA's Lunar Reconnaissance Orbiter. Credit: NASA LRO WAC science team

LROC isn't just constantly taking pictures of whatever is underneath it. Instead, the lunar surface is continuously being imaged based on targets that have been chosen by planetary scientists to answer specific science questions. While planning imaging observations, several things need to be accounted for, such as lighting conditions, temperatures, angles and timing. Deciding on targets is a team effort, said Lillian Ostrach, who worked with the LROC science team as a graduate student at ASU and is now at Goddard as an LROC science team member. "There's an international science team that provides input and the operations team that helps identify locations on the Moon that should be imaged," she said. "Everyone on the team supports targeting, image acquisition and image processing."

When the mission launched, one of the primary goals for LROC was to study the surface at meter-scale resolution for potential future landing sites. That included geologically "exciting" areas on the Moon that were also safe landing sites.

But LROC's main goal now is to help tell the geologic story of the Moon by looking at the properties of the lunar regolith, mapping the mineralogy of the lunar surface with WAC, creating the detailed digital terrain models by combining high-resolution NAC stereo observations with data from the LOLA, and determining the current impact rate.

One surprise found by analyzing LROC images has been the evidence of very young volcanism. Previously, scientists estimated that lunar volcanism ended about a billion years ago. But LROC images have helped researchers better date the Moon's volcanic activity, and understand that it slowed gradually instead of stopping abruptly. Scores of distinctive rock deposits observed by LROC are estimated to be less than 100 million years old, and some areas may be less than 50 million years old.

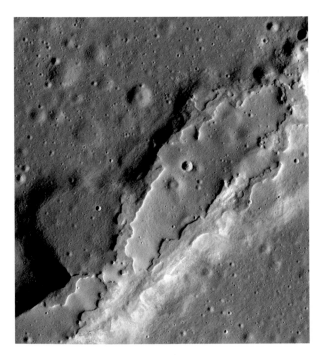

Lava flows have spread across the floor of this large collapsed area. Their lack of impact craters and steep sides show that they erupted relatively recently. Credit: NASA/GSFC/ Arizona State University

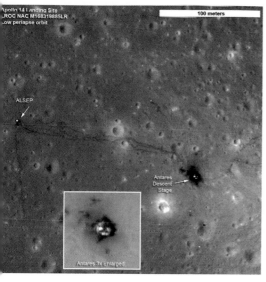

The paths left by astronauts Alan Shepard and Edgar Mitchell on both Apollo 14 moon walks are visible in this image from LRO's lowest orbit. (At the end of the second moon walk, Shepard famously hit two golf balls.) The descent stage of the lunar module Antares is also visible. Credit: NASA's Goddard Space Flight Center/ASU

The Apollo landing sites have always been of particular interest and at one point in the mission, LRO's orbit was lowered to just 13 miles (21 km) over portions of the Moon's surface. This very low orbit provided the opportunity to have sharper views of the historic Apollo sites.

"To me the later pictures were equally exciting if not more so," Robinson said. "We were down lower and the Sun was higher above the horizon, thus the tracks and hardware were much sharper. Also we obtained images many times, each time with a different Sun angle. The comparison of the surface at varying illumination really allowed a detailed look at the sites and provided the means to reinterpret the geologic context of each sample station."

For example, by looking at the series of images from the Apollo landing sites, the team could tell the flags put in place by the astronauts were still standing because of the shadows they created, except for the Apollo 11 flag, which fell over.

"Buzz Aldrin reported that he saw the Apollo 11 flag fall down when hit by the LM exhaust," Vondrak said, "and this is verified by the LRO images, as well as video taken from the LM. The LRO images show the shadows of the flags at the other sites, but the flags have probably faded and deteriorated due to radiation and the strong solar ultraviolet light."

Vondrak said LROC images have also helped give context to the Moon rocks brought back by the Apollo astronauts, as well as help solve a few mysteries.

"One of the big puzzles of Apollo 14 with Edgar Mitchell and Alan Shepherd," said Vondrak, "was where the rocks came from when they went to a nearby crater called Cone Crater. They never made it to the rim of the crater, and they never knew how close they got. And the locations of where they picked up the rocks were not well documented. They took pictures but people couldn't locate those features in previous orbital data. So one geological puzzle was, what was the provenance of the Apollo 14 rocks?"

LROC's sharp eyes could see the trail of where the astronauts walked, and images showed the trail to the boulder stations where the astronauts stopped. "So now we know very well where these rocks came from," Vondrak said, "and we also know the astronauts could have made it to the crater rim if they had only walked another 100 feet (30 m). They accomplished their science goals by bringing back rocks from near the rim, but they just missed out on seeing the spectacular view across Cone Crater."

Apollo 17 astronaut Harrison (Jack) Schmitt, the only geologist who went to the Moon, has been looking at orbital data from LROC, as there was a question of what boundaries he and his crewmate Gene Cernan sampled rocks from. Schmitt is working with Robinson and Petro, and they have written a scientific paper about their finding.

"It reexamines the Apollo 17 landing sites in light of the LROC data and a few other recent data sets," said Petro. "What Jack has observed and I think we all agree, the age of the surface around landing site is different than we've thought for the past 45 years. We didn't have the whole story, but LROC has provided new information, and that is calling into question some of the conclusions and assumptions that were made about the landing site. That's been incredible."

But all the LROC images highlight that the Moon is a fascinating place. In 2016, pictures from LROC were part of a special display called *A New Moon Rises* at the Smithsonian's National Air and Space Museum in Washington DC. I had the chance to see the exhibit, and it was awe-inspiring to view the huge wall-size panoramas from LROC showing the Moon in intricate detail. Even minor variations in the lunar surface are visible in the large images, revealing even more of the Moon's stark beauty. You can see the images for yourself on the LROC website (http://lroc.sese.asu.edu/) where more than a million images are in the LROC archive.

Robinson said he hopes the beauty of the images will perhaps renew interest in the Moon, with the images of the Apollo mission sites perhaps sparking the aspirations for a return. "To me the LROC images reveal the Moon to be a mysterious and beautiful place—a whole world just three days away," he said.

THE FUTURE OF THE MOON

It remains to be seen how long LRO's mission will continue at the Moon and whether the orbiter will be joined by future missions and future explorers. Two other recent NASA robotic missions to the Moon include the twin GRAIL spacecraft that measured the Moon's gravity to extremely fine detail, but ended in 2012, and the Lunar Atmosphere and Dust Environment Explorer (LADEE), which gathered detailed information about the lunar atmosphere before ending in 2014.

LRO captured a unique view of Earth from the spacecraft's vantage point in orbit around the Moon that evokes the famous "Blue Marble" image taken by the Apollo astronauts. Credit: NASA/GSFC/Arizona State University

Noah Petro. Credit: NASA/GSFC

The hope of all the scientists and engineers I talked to for this book is that NASA's robotic exploration can continue in a robust manner, and that the space agency can design a sustainable and meaningful exploration program that includes both robotic and human missions working together, all to spur innovation and technology. That benefits everyone on Earth, as well as feeding the need to explore and push the boundaries of our limits, scientifically, technologically and philosophically.

And some consider the Moon a "gateway" to be able to explore the rest of the Solar System. Besides the interest from governmental space agencies there is commercial interest in lunar exploration as well, including the Google Lunar XPRIZE, a $30 million prize competition with teams contending to land a privately funded rover on the Moon, travel 1,640 feet (500 m), and transmit back high-definition video and images. In the current schedule, teams have until the end of 2016 to announce a verified launch contract to remain in the competition and complete their mission by the end of 2017.

"I'd love to see us go to Mars," Vondrak said, "I hoped I would be alive to applaud at my TV when astronauts landed on Mars. But Mars is far away, it's very expensive, it's also very risky. But the Moon is our nearest neighbor, it's closer, and now with LRO we know the Moon very well and we know where we can go safely and where there are resources. With Apollo we visited the Moon, and the next step is to learn how to live there and to work there. Living and working on the Moon will prepare us to go to Mars and beyond."

While the spacecraft like LRO continue on their journeys of exploration, humans are the fuel that enables exploration.

"It's the people ultimately that make the missions go," said Petro. "For LRO, all the instrument teams and the folks at Goddard that implement what the teams want to do, without them, we don't have a mission. We'd just have a hunk of metal orbiting the Moon. I'm so grateful to be part of this mission and for the opportunity to work with this team. It's a really special group of people."

CHAPTER 10

SEEING THE FUTURE: MISSIONS AND DISCOVERIES TO WATCH

TOW TRUCKS IN SPACE

"What I'm showing you here today is the future," said Frank Cepollina as he nimbly guided me around the Robotic Operations Center at the Satellite Servicing Capabilities Office (SSCO) at NASA Goddard. "But this isn't the final piece of the future, just a step in the path of the future."

This impressive research facility contains several different areas where industrial robotic arms of various sizes stand poised to work on mock-ups of satellites. Technicians tinkered with one arm, while specialists called **robos** sat at computer workstations. Looming large on one side of the expansive facility was a model of a five meter boulder from an asteroid being grappled by huge metallic legs.

Cepollina, both a legend and a warm, approachable father figure, made me feel like an honored guest at what he called "his" lab. "I want to show you all my toys," he smiled. All the technicians and specialists stopped whatever they were doing to affectionately greet "Cepi."

The Robotic Operations Center (ROC) is a test bed for satellite-servicing technologies that enable science and exploration. Credit: NASA/GSFC/SSCO

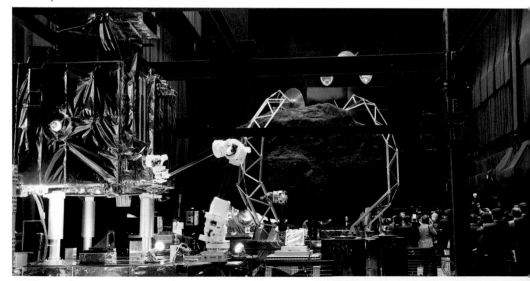

Artist's concept of Restore-L, a robotic spacecraft equipped with the tools, technologies and techniques needed to extend satellites' lifespans - even if they were not designed to be serviced in orbit. Credit: NASA

This giant lab is where Cepi's kids, as he calls them (no matter their ages), test technologies and operations for numerous types of missions, including robotic refueling and repair missions to satellites in orbit. They also test equipment for NASA's proposed Asteroid Redirect Mission, where a robotic craft would collect a multiton boulder from an asteroid's surface and redirect it into a stable orbit around the Moon.

Cepollina, as you recall from Chapter 3, led the Hubble servicing missions. Now he is the associate director of the SSCO, and he helped develop a new mission called Restore-L, which is a robotic spacecraft that is equipped with the tools and technologies needed to extend the life span of satellites, even those that were not designed to be serviced in orbit.

"When the space shuttle was taken out of service," Cepollina explained, "the next thing we focused on was how we could do in-orbit tasks like maintenance or scientific exploration but without astronauts and the shuttle. Our whole evolution was to create a robot that would be able to act like a tow truck but without a driver. And it could go to satellites that are sick—ones that need repair, or if an antenna didn't deploy correctly, or they need more fuel."

The SSCO has developed smart software that will enable the "tow truck" to autonomously capture an uncontrolled spacecraft or a rotating asteroid.

"Even though a spacecraft or asteroid might be tumbling, turning and rotating," Cepollina said, "the robot arms on our tow truck are smart enough to be able to see what the object is doing and calculate the spin rate and grab it."

Frank Cepollina with a Motoman robot arm in the ROC. Credit: Nancy Atkinson

While the approach and capture of objects in space is autonomous, the beauty of this system, Cepollina said, is that operators on Earth can manipulate the robotic arm to conduct exquisitely detailed work. In the case of fixing satellites, the team has devised operations to work on spacecraft, no matter their configuration.

To make this possible, NASA has worked with the engineers who developed the da Vinci surgical robot that performs intricate teleoperated surgery, from tiny, computer-controlled incisions for heart surgery or removing cancerous tumors. Both the surgical and satellite robots use a 3-D visualization system to guide their operations from afar.

The ROC facility has black, curtain-lined walls and when all the lights are shut off, it simulates the darkness of space.

"We have artificial lights that produce the glows, glints and glimmers of space," Cepollina explained, "and so this room becomes a fully immersive training facility, with full-scale mock-ups of parts of several different satellites so our robos can practice in reasonable fidelity, seeing in their computer screens the same conditions of what they'd be seeing as our tow truck is working on a satellite in space. We practice, practice, practice to make sure the degrees of sophistication we are trying to incorporate are eventually easily attainable."

One disadvantage of teleoperated robots is they work slower than a human on location would.

"But they don't have to eat or sleep." Cepollina smiled. "And we can have the robo operators take shifts on Earth."

What satellites could be fixed?

"There are billions of dollars of assets in high geosynchronous orbits," said Cepollina, orbits where satellites are about 22,236 miles (35,786 km) above Earth. This orbit provides a perfect position for about 400 weather, communications and surveillance satellites. "NASA, NOAA, the military and commercial satellites are there, where the shuttle couldn't go. There are plenty of potential clients."

Cameras, monitors and computer stations are the eyes and brains of the facility, giving robot operators like Joe Easley a front-seat view of the robots' activities as they issue commands. Credit: NASA/GSFC/SSCO

Robots and mockups at the Robotic Operations Center at the Satellite Servicing Capabilities Office at Goddard Space Flight Center help simulate and practice satellite-servicing tasks, such as gently grasping a client satellite. Credit. NASA/GSFC/SSCO

Restore-L is scheduled to launch in 2019 and it will rendezvous with, grasp, refuel and relocate a currently unnamed government-owned satellite to extend its life. With a successful mission, the hope is that servicing technologies could be incorporated into other NASA missions, including exploration and science ventures. While this technology is being developed by NASA, the hope is that Restore-L will jumpstart a domestic satellite servicing industry.

This has been Cepollina's dream from the beginning.

"To me, it was astounding that we would just throw satellites away on orbit," he said. "It seemed we should find a way to fix these satellites for economic reasons, and for the scientific benefits we could derive. I wanted to find a way to fix and upgrade satellites."

With the space shuttle, Cepollina and his team conducted six servicing missions (five to Hubble and the Solar Max satellite repair). So while this isn't a new idea, the SSCO is working to develop the sophisticated teleoperated technology needed to ensure success.

"With the shuttle now gone, we took our greatest experiences working with astronauts and converted those tools to work with robots," Cepollina said. "Our hope is to evolve to the point where, like Hubble, we'll be able to get a satellite to work for many, many more years."

Since 2011, the International Space Station has housed the Robotic Refueling Mission experiment, demonstrating robotic satellite-servicing tools, technologies and techniques developed by the SSCO. There, robotic operators at Johnson Space Center in Houston operate the Canadian-built robotic arm and the twin-armed large robot called Dextre, both on the exterior of the International Space Station. Additional experiments will be sent to the space station to test out more robotically operated missions. Again, the idea is to eventually service satellites and then turn the technology over to commercial companies who can build this into a business.

Launched in 2011, the Juno spacecraft arrived at Jupiter in July of 2016 to study the giant planet from an elliptical, polar orbit. Juno dives repeatedly between the planet and its intense belts of charged particle radiation, traveling from pole to pole in about an hour, and coming within 3,000 miles (5,000 km) of Jupiter's cloud tops at closest approach. Credit: NASA/JPL-Caltech

"We're doing all kinds of things, and the whole essence of what we're doing is trying to evolve the capability of producing new technology and new knowledge of the universe," Cepollina said.

Even at age 80, Cepollina still dreams like a kid. The final servicing mission to Hubble sparked an idea for how a large telescope could be constructed in Earth's orbit.

"The Museo Galileo (the Galileo Museum) in Florence, Italy, built an exact replica of Galileo's original 400-year-old telescope to be sent along on the servicing mission," Cepollina recalled. "When all the repairs were done, the astronauts took it out, laid it across the aft flight deck of the shuttle and they took a picture. Visible out the window was Hubble. I said, 'there's a message there: in 400 years, we went from 1 inch (25 mm) to 100 inches (254 cm). That's a poor transition. Time for us to move ahead.'"

Cepollina's concept is a 1,000-inch (2,540 cm) telescope with giant mirror segments assembled by robots and humans together in space.

"We could really stretch our ability and look for the potential of other life in our galaxy," Cepollina said passionately. "That's my ultimate goal, to stretch the imagination of our people. I see this as a remarriage of robotics and astronauts all over again, where robots do the heavy lifting and humans do the heavy thinking. When you combine the two, you've got something really special."

ON THE HORIZON

Let's look at several new and upcoming missions, all unique and exciting robotic explorations of the cosmos.

STUDYING JUPITER UP CLOSE: JUNO

Launch: August 2011
Arrival: July 2016
Mission End: February 2018

Jupiter is the most massive planet in our solar system, and with its menagerie of moons and an enormous magnetic field, this unique world is like a miniature solar system. Scientists have said understanding Jupiter and how it formed would help answer many questions not only about this giant planet, but also about how the solar system as a whole came together.

This artist's rendering shows NASA's Juno spacecraft making one of its close passes over Jupiter. Credits: NASA/ JPL-Caltech

A new mission to Jupiter began in July of 2016 with the arrival of the Juno spacecraft. Juno is unlocking the mysteries of this giant planet and is investigating Jupiter's origins, its interior structure, its atmosphere and magnetosphere from an unusual and innovative orbit with a suite of seven science instruments. In addition, a camera called JunoCam will be used by students and the public to take pictures of the gas giant, including the first-ever images of Jupiter's polar region.

"Juno is a combination of several innovative science and mission design concepts," said Juno's project manager, Rick Nybakken, at JPL. "This is the best-designed and most elegant mission I've ever been a part of."

Nybakken said that when he read the proposal for the mission, he thought it was breathtaking. From the unprecedented up close investigations of Jupiter to its innovative use of solar panels, this spinning spacecraft brings a full payload of instruments to study the gas giant planet as never before.

Juno engineers designed the mission to enable the use of solar panels, which prior to Juno have never been used on a spacecraft going so far from the Sun. Juno orbits Jupiter with its solar panels always pointed toward the Sun, and it never goes behind the planet. Juno's orbital design not only enabled a historic solar-powered mission, it also established Juno's unique science orbit.

"Juno has an elliptical orbit that brings it between the inner edges of Jupiter's radiation belt and the planet at only 3,107 miles (5,000 km) above the cloud tops," said Nybakken. "This close proximity to Jupiter is unprecedented, as no other mission has conducted their science mission this close to the planet. We're right on top of Jupiter, so to speak."

Why so close to a planet that is known for its harsh radiation?

"To make Juno's groundbreaking science measurements, we have to operate closer to Jupiter than any previous mission," Nybakken said. "While there is risk in operating in uncharted territory, we've spent a lot of time thinking about what the risks are and how to minimize them."

It will take fourteen Earth days for Juno to complete an orbit of Jupiter, and the orbit the mission planners designed minimizes the radiation the spacecraft receives, at least early in the mission. Additionally, the majority of the electronics for the spacecraft are placed in a special titanium radiation vault for protection.

"Protecting our electronics in a radiation vault significantly reduced the redesign that is typically needed to operate in Jupiter's intense radiation environment," Nybakken said. "Less redesign freed up more funds for Juno's powerful science payload. I think of Juno as a scientific tour de force, as a lot of innovative concepts were merged into a scientifically groundbreaking yet very cohesive mission."

Special instruments are studying Jupiter's radiation belt and magnetosphere, its interior structure and the turbulent atmosphere, as well as providing views of the planet with spectacular close-up images.

"Juno is really a data-driven mission," said Juno project scientist Steve Levin, also at JPL, "with the driving force of getting the data while we can."

Juno will be able to orbit Jupiter for only about twenty months (approximately 37 orbits) because the radiation environment will eventually make the spacecraft inoperable.

"Jupiter presents a lot of problems as far as being nice to instruments," Levin said. "But that's the trade-off of being able to make the measurements we want."

The data Levin anticipates most is the amount of the global water abundance at Jupiter. This data is an unanswered mystery from the only other mission that orbited Jupiter, the Galileo spacecraft, which was at Jupiter from 1995 to 2003. The mission made many new discoveries, but it also left some unanswered questions. Part of the mission was a probe that descended into Jupiter's atmosphere.

"The probe found hardly any water at all," Levin said, "and that's a mystery that is waiting to be solved, because knowing how much water is at Jupiter is key to understanding how the planet formed. It's possible the probe went to one location where there was no water. So Juno will use a microwave instrument that can measure the amount of water for the entire atmosphere instead of just measuring it in one place or just a few places. That is a really big advantage."

Other instruments will study the magnetic field, the planet's interior and more. Levin said he's looking forward to seeing something that's never been seen before: Jupiter's poles.

"It will be nice to know what the poles look like," he said, "and I'm excited for what we're doing with the visible light camera. We're making JunoCam an instrument that belongs to the public, as much as we possibly can. We'll solicit the aid of the public in picking where to look and releasing the images as quickly as we can."

Mission images are available at www.missionjuno.swri.edu/media-gallery/junocam.

At the end of the mission, projected to be in February of 2018, Juno will be deorbited into Jupiter to avoid potential contamination of Jupiter's moon Europa, thought to potentially be a habitable environment. Like the Cassini mission, conducting a controlled deorbit is a requirement of NASA's Planetary Protection guidelines.

Artist concept of the ExoMars/Trace Gas Orbiter mission. Credit: NASA/ESA

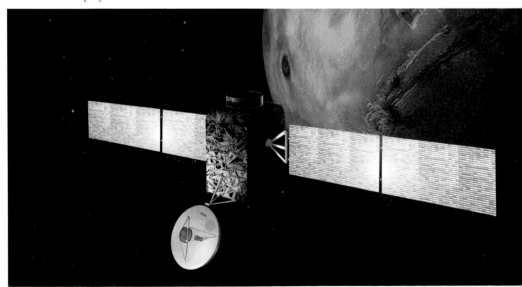

The OSIRIS-REx spacecraft being lifted into the thermal vacuum chamber at Lockheed Martin in Colorado for environmental testing. Credit: Lockheed Martin

THE NEXT MISSION TO MARS: EXOMARS

Launch: March 2016
Arrival at Mars: October 2016

The ExoMars 2016 mission is a joint mission between ESA and Russia (Roscosmos), which is sending a pair of European-built spacecraft to study Mars: the Trace Gas Orbiter and Schiaparelli, an entry, descent and landing demonstrator module. The orbiter is scanning for signatures of methane gas that could possibly be an indication of life or of nonbiological geological processes taking place on Mars today. Telescopes on Earth as well as the Curiosity rover have detected the mysterious presence of methane, and this mission hopes to provide more insight into why this gas that quickly decays is seemingly being produced on Mars. The orbiter will also study Mars' environment, look for signatures of water and study the planet's surface.

The lander is a test of landing large payloads on Mars but will also collect samples with a drill and test out other capabilities for future ESA missions.

This mission was originally a joint venture between ESA and NASA, but the United States backed out in 2012 due to budget concerns.

Current plans are to follow this mission in 2018 with a Roscosmos-built lander (ExoMars 2018 surface platform) that will deliver the ESA-built ExoMars Rover to study Mars' surface.

OSIRIS-REX: MISSION TO AN ASTEROID

Launch: September 2016
Target: Asteroid Bennu
Arrival: October 2018
Sample Return to Earth: 2023

The OSIRIS-REx mission looks to help answer all those questions that humans have been asking for centuries: Where did we come from? What is our destiny? Scientists are hoping an asteroid named Bennu will help provide some answers.

Launched in September of 2016, the spacecraft is expected to reach its target in 2018, study the asteroid for about one year, collect a sample and then return back to Earth in 2023.

Asteroids represent the leftover debris from the formation of our solar system over four billion years ago, and these ancient bodies can teach us about the history of the Sun and planets. Scientists think Bennu, with a diameter of 538 yards (492 m), may contain the molecular precursors to the origin of life, as well as water and precious metals.

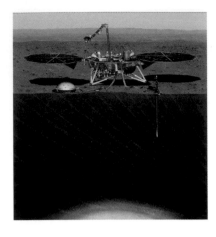

This artist's concept from August 2015 depicts NASA's InSight Mars lander fully deployed for studying the deep interior of Mars. Credit: NASA/JPL-Caltech

However, Bennu is also considered one of the most potentially hazardous asteroids, as it has a relatively high probability of impacting Earth late in the twenty-second century. OSIRIS-REx will determine Bennu's physical and chemical properties, critical for scientists attempting to develop methods to mitigate the probably of its impact with Earth, perhaps with a gravity-tug or an impact mission to Bennu.

DELAYED MARS LANDER: INSIGHT

Launch: May 2018
Landing on Mars: November 2018

Originally planned for launch in March 2016, the Interior Exploration using Seismic Investigations, Geodesy and Heat Transport (InSight) mission had to be delayed for two years due to a leak in one of the main instruments. InSight will land on Mars to study the deep interior of the red planet.

InSight's primary goal is to help us understand how rocky planets—including Earth—formed and evolved, and the lander is based on the successful Phoenix mission to Mars that took place in 2008. The lander has a drill, seismometer and heat transfer probe to study Mars' surface and interior, and provide insight on the planet's early geological evolution. This is an international mission with researchers from Austria, Belgium, Canada, France, Germany, Japan, Poland, Spain, Switzerland, the United Kingdom and the United States.

Elements of the BepiColombo Mercury Composite Spacecraft. From left to right: Mercury Transfer Module (MTM), Mercury Planetary Orbiter (MPO), Magnetospheric Orbiter Sunshield and Interface Structure (MOSIF) and Mercury Magnetospheric Orbiter (MMO). Credit: ESA

BEPICOLOMBO: EUROPE'S FIRST MISSION TO MERCURY

Launch: 2018
Arrival: 2024

BepiColombo is Europe's first-ever mission to Mercury. Set to launch in 2018, it should arrive at Mercury—which some call the least explored terrestrial planet in our Solar System—in late 2024. The mission is actually two spacecraft: the Mercury Planetary Orbiter (MPO) which will study the surface and internal composition of the planet, and the Mercury Magnetospheric Orbiter (MMO)

Standing tall and glimmering gold inside NASA's Goddard Space Flight Center's clean room in Greenbelt, Maryland is the James Webb Space Telescope primary mirror. It is considered the biggest and most powerful space telescope ever launched. Credit: NASA/Chris Gunn

will study the region in space around Mercury that is influenced by the planet's magnetic field. BepiColombo is a joint mission between ESA and the Japan Aerospace Exploration Agency (JAXA), and it is scheduled to last about 2 years.

THE NEXT GENERATION OF SPACE OBSERVATORIES: JAMES WEBB SPACE TELESCOPE

Launch: October 2018
Mission Duration: Five to ten Years
Location: One million miles (1.5 million km) from Earth

The James Webb Space Telescope (JWST) is the much-anticipated, long-awaited next-generation telescope. Planned for launch in October 2018, JWST has been touted as the successor to the Hubble Space Telescope. With it, astronomers hope to look back in time to when the universe was just 200 million years old and see the first stars and galaxies.

JWST is a large infrared telescope with a gold-coated 21-foot (6.5-m) primary mirror (gold is an excellent reflector of infrared light, as opposed to the aluminum that coats the Hubble Space Telescope Mirror). An international mission, the telescope will be launched on an Ariane 5 rocket from French Guiana. JWST will be the premier observatory of the next decade, serving thousands of astronomers worldwide. It will study every phase in the history of our universe, ranging from the first luminous glows after the big bang, to the formation of solar systems capable of supporting life on planets like Earth, to the evolution of our own solar system. JWST will be a general purpose observatory, meaning astronomers can use it to study the "hot topics" that might arise, much like Hubble. In fact, JWST will be run just like Hubble in that astronomers can request time to use the observatory.

It will be located one million miles (1.5 million km) away from Earth at what is called the second Lagrange point, or L2, a place out past the Moon, where gravity from the Earth, Moon and Sun provide a stable place to hang out. However, that means it can't be repaired or serviced like Hubble was. It will take JWST about a month to reach its orbit and should reach full operations in about six months.

Several innovative technologies were developed for JWST, including a primary mirror made of eighteen separate segments that unfold and adjust to shape after launch, a tennis court-size five-layer sun shield to protect it from excess heat and light, and four instruments—cameras and spectrometers—that can detect extremely faint and distant signals.

JWST has been a long time coming. First proposed in the early 1990s, the mission was originally approved in 1996. Like Hubble, delays and cost overruns have occurred due to the construction of unique components and systems. Launch dates slipped from 2007, 2011, 2013 with a supposedly firm launch date now of 2018. The mirror was fully assembled in early 2016. The total cost to build JWST has gone from an original estimate of about $1 billion to now being about $8 billion.

Since it is not an optical telescope like Hubble, the images produced by JWST will be somewhat different. But hopes are high for what this observatory will be able to tell us about the universe.

MARS AGAIN: MARS 2020 ROVER

Scheduled Launch and arrival: 2020

This mission is based on the very successful Curiosity rover, which landed on Mars in 2012. It will use the same body chassis as Curiosity but with new science instruments for new science objectives. After pulling out of the ExoMars mission with ESA, NASA decided to go to Mars on its own (however, there will be international participation from Spain, France and Norway).

Diagram of the proposed science instruments for the Mars 2020 rover, with notations of the countries that are contributing the instruments. Credit: NASA/JPL

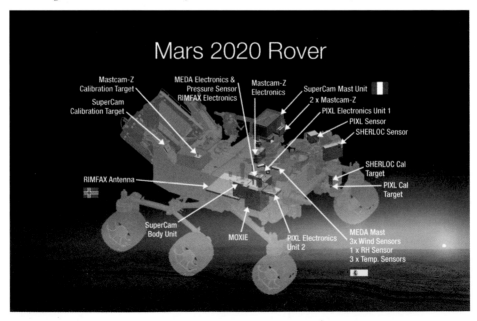

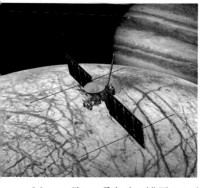

An artist's rendering of the proposed mission to study Europa. Credit: NASA/JPL

Plans for the Mars 2020 rover are to identify and select a collection of rock and soil samples that will be stored for potential return to Earth by a future mission, search for the signatures of past life on Mars, and test ways for future human explorers to use the resources available on Mars to "live off the land." This includes understanding the hazards posed by Martian dust and demonstrating technologies to process carbon dioxide from the atmosphere to produce oxygen, which could be used for the production of fuel. The rover is expected to have several updated cameras, upgraded hardware and new instruments to conduct geological assessments of the rover's landing site, determine the potential habitability of the environment, and directly search for signs of ancient Martian life.

Sarah Milkovich, Science Systems Engineer for the Mars 2020 Rover said one goal in preparing for this mission is to make the rover easy to operate. "We've got very ambitious goals for this rover," she said, "and so we want to automate processes where we can and to enable the rover to make more decisions, allowing the operations team to focus its attention on the decision-making that only a human can do."

MISSION TO JUPITER'S MOON EUROPA

Launch: TBD (mid-2020s)

An orbiter to Europa will conduct detailed reconnaissance of this enticing moon to investigate whether the icy moon could harbor conditions suitable for life. Thought to have a global subsurface ocean beneath its icy crust, scientists have long speculated that conditions there could be favorable for life.

The plan is to send a radiation-tolerant spacecraft into a long, looping orbit around Jupiter to perform repeated close flybys of Europa. Nine science instruments have already been selected, including cameras and spectrometers to produce high-resolution images of Europa's surface and determine its composition. An ice-penetrating radar will determine the thickness of the moon's icy shell and search for subsurface lakes similar to those beneath Antarctica's ice sheet. The mission will also carry a magnetometer to measure the strength and direction of the moon's magnetic field, which will allow scientists to determine the depth and salinity of its ocean.

A thermal instrument will survey Europa's frozen surface in search of recent eruptions of warmer water at or near the surface, while additional instruments will search for evidence of water and tiny particles in the moon's thin atmosphere. The Hubble Space Telescope observed water vapor above the south pole region of Europa in 2012, providing potential evidence of water plumes. If the plumes' existence is confirmed—and if they're linked to a subsurface ocean—studying their composition will help scientists investigate the chemical makeup of Europa's potentially habitable environment while minimizing the need to drill through layers of ice.

WHY EXPLORE SPACE?

As mentioned in the introduction, this book provides a snapshot in time of stories about robotic missions to space. As is inevitable in journeys of exploration, things are always changing because of new discoveries or unpredicted developments. As a space journalist, I've always known this to be true, but it was strikingly confirmed while writing this book. I had to go back and rewrite parts of several chapters because of exciting new findings or potential big changes for missions. For example, the Kepler mission announced in May 2016 they had verified 1,284 new planets—the single largest finding of planets to date, and just a few weeks earlier, the Dawn mission team proposed the totally unexpected concept of exploring an additional target.

But there's one thing that will likely never change: the dedication and commitment of the people who work on these missions to explore the cosmos. I felt honored and humbled to talk with the thirty-five scientists and engineers who shared their stories for this book, and time and time again, I was amazed by their enthusiasm and excitement for what they do. I truly hope I was able to capture and convey their incredible spirit of exploration and discovery.

But, just why do we explore space anyway? Back when early explorers like Magellan, da Gama or even Pythéas set off on their journeys, undoubtedly many people thought it was foolish to risk lives and spend a lot of money to find out what was beyond the horizon. But those explorers didn't find what they expected, and their explorations ended up changing the world. Similarly, robotic space missions and telescopes, as well as ground-based telescopes have helped us explore remotely and have facilitated the discovery of so many things we didn't know—and didn't expect—about our universe.

Part of what drives us to explore and discover is the intangible: expanding our horizons, feeding our curiosity, finding all those unexpected things and trying to answer those profound questions discussed in previous chapters, like how did the universe begin? How did life begin? Are we alone?

Then there are the tangible benefits of exploration, such as the development of technology and advances in science, medicine, communications, transportation and much more. The spirit of exploration inspires us to create and invent so that we can go explore and in turn, possibly change the world. We don't know yet exactly what we'll find if humans ever go to Mars, Europa or beyond, but if we stay in our earthly "cradle," we'll never find out.

But exploration takes money and the most often-used argument against space exploration is that we should use that money to alleviate problems here on Earth. But that argument fails to realize that NASA and other space agencies around the world don't just pack millions of dollar

With giant Saturn hanging in the blackness and sheltering Cassini from the Sun's blinding glare, the spacecraft viewed the rings as never before, revealing previously unknown faint rings. This marvelous panoramic view was created by combining a total of 165 images taken by the Cassini wide-angle camera over nearly three hours on Sept. 15, 2006. The color contrast is exaggerated in this view by digitally compositing ultraviolet, infrared and clear filter images and was then adjusted to resemble natural color. The white dot just inside the second outermost ring is planet Earth. Credit: NASA/JPL/Space Science Institute.

bills into a rocket and blast them into space. All that money is spent on Earth, providing jobs and the opportunity for some of the world's brightest minds to use their talents to benefit humanity and very likely change the world for the better. The exploration of space spurs inventions that we use every day, many which save lives and improve the quality of life, and the technologies developed for space can then be used to create new companies and industries, leading to more jobs and opportunities. NASA and other space agencies provide that all-important research and development that every successful company thrives on.

Public opinion polls show that many Americans have a misconception of how much money NASA receives from the United States government. NASA's proposed budget for fiscal year 2017 is $19 billion. That sounds like a lot of money, and it is, but to put it in perspective, the total proposed 2017 federal budget is $4.2 trillion. That means NASA's cut of the U.S. budget is less than a half of one percent (0.45%), or less than half a penny of every tax dollar. By comparison, the proposed 2017 budget for education is 2% of the U.S. budget ($85 billion), military is 15% ($632 billion) and Social Security, unemployment and labor is 33% ($1.39 trillion).

But space exploration doesn't just help us learn about the cosmos, it also helps us learn about our own world and ourselves. In a rather famous letter from 1970, Dr. Ernst Stuhlinger, then-associate director of science at NASA's Marshall Space Flight Center, responded to a letter he had received which asked how NASA could suggest spending billions of dollars on going to Mars when so many children were starving on Earth.

"Although our space program seems to lead us away from our Earth and out toward the moon, the sun, the planets and the stars, I believe that none of these celestial objects will find as much attention and study by space scientists as our Earth," Stuhlinger wrote. "It will become a better Earth, not only because of all the new technological and scientific knowledge which we will apply to the betterment of life, but also because we are developing a far deeper appreciation of our Earth, of life, and of man."

Looking both outward and inward—and living life with a sense of wonder—can help us create a better world right now and build a better world for future generations.

ACKNOWLEDGMENTS

A sincere debt of gratitude to all the NASA scientists and engineers who took time out of their extremely busy lives to visit with me, give me tours, answer follow-up questions and review technical details. Their kind acceptance and enthusiasm for this project was overwhelming and I can't overstate how much their efforts mean to me, both personally and as a writer. At the Jet Propulsion Laboratory: Ashwin Vasavada, John Michael Morookian, Marc Rayman, Keri Bean, Linda Spilker, Earl Maize, Rich Zurek, Dan Johnston, Wesley Traub, Rick Nybakken, Steve Levin and Neil Mottinger. At Goddard Space Flight Center: Dean Pesnell, Alex Young, Rich Vondrak, and Frank Cepollina. At the Johns Hopkins Applied Physics Lab: Hal Weaver and Alice Bowman. At the Space Telescope Science Institute: Ken Sembach, Zolt Levay, Helmut Jenkner and Carol Christian. Phone and email interviews include Alan Stern (Southwest Research Institute), Mark Robinson (ASU); Alfred McEwen, Christian Schaller, Kristin Block and Ari Espinoza, (University of Arizona); Robert West and Sarah Milkovich (JPL), Natalie Batalha, Tom Barclay, Tony Colaprete and Jennifer Heldmann (NASA Ames Research Center), Noah Petro and Lillian Ostrach (Goddard), and Tom Woods (Laboratory for Atmospheric and Space Physics). Again, thank you all for sharing your stories and experiences, and providing a window into the amazing work you do.

Heartfelt extra thanks to Marc Rayman for his extended discussions on ion propulsion and to Noah Petro and Rich Vondrak for their extended discussions on LRO.

Special thanks to the JPL media office, especially Mark Petrovich, Guy Webster, DC Agle and Elizabeth Landau for coordinating interviews and shepherding me through the beautiful JPL campus. Being there was a dream come true. Another big thank you to all the folks at the NASA Goddard Office of Communications, especially Karen Fox for helping to coordinate my visit, as well as Sarah Frazier, Nancy Neal Jones, DeWayne Washington and Adrienne Alessondro for deftly escorting a "risky" Minnesotan around. Special thanks to Michael Buckley for arranging the wonderful, memorable visit at Johns Hopkins APL, and incredible thanks to Cheryl Gundy, for the amazing array of tours and interviews arranged at the iconic STScI. Thanks for making me feel like a VIP! Also thanks to Ray Villard at STScI for the chance to meet and chat about science writing. A special thank you to Michele Johnson at NASA Ames for arranging interviews for Kepler.

Thanks also to Jim McClure at JPL for the tour of the Space Flight Operations Facility and to Jim Wang for showing me around the Mars Yard. A shout-out to Aries Keck for including me in the NASA Social event at Goddard and everyone there who provided tours and insights.

Also to my friends at JPL: thanks to Neil Mottinger for our long-time friendship, for the special tour, being a lunch buddy and for becoming part of this book; to Chris Potts for squeezing in the chance to finally meet in person; to Kay Ferrari of the Solar System Ambassador Program for being such a wonderful person.

As always, never-ending thanks and appreciation to my mentor, boss and friend, Universe Today's founder and publisher Fraser Cain. Thank you for taking a chance on an untested writer all those years ago, giving me the opportunity to 'ride along' on all the missions and space journeys. You're the best! Other important mentors along my path include Pamela Gay, Phil Plait, Ian O'Neill, the National Association of Science Writers and the Council for the Advancement of Science Writing, and all the wonderful writers I've worked with at Universe Today since 2004. At the risk of leaving someone out, I do want to recognize my longtime co-writers Jason Major, David Dickinson, Ken Kremer, Elizabeth Howell, Matt Williams and the late Tammy Plotner. And special extra heartfelt thanks to UT writer Bob King and author of Page Street's *Night Sky for the Naked Eye* for his encouragement and companionship in our "parallel universes" as we wrote our manuscripts. Skål!

While most of the images used in this book were taken by the spacecraft out there exploring our solar system and universe, I do want to recognize the talented NASA photographers who excel at capturing the spectacular launches and behind-the-scenes shots, as well as capturing the personalities behind the missions. In particular, many thanks to the remarkable Bill Ingalls, who took several of the images used here. Also my appreciation to all the NASA and ESA graphic artists for their beautiful renditions of the cosmos and excellent graphics that help explain the complexities of space exploration. Special thanks to graphic artist Kevin Gill who provided his graphic on comparative sizes of KBOs, to Jason Major for his rendition of Curiosity's photobombing selfie from sols 612-613, and to Bob King for his beautiful aurora image.

To everyone at Page Street: Wow, you are all absolutely wonderful! I'll never forget the day I was contacted by publisher Will Kiester and editor Elizabeth Seise. Thank you, Will, for this incredible opportunity, and Elizabeth, thank you for guiding me through every step of this process; you were absolutely awesome to work with! Special thanks to copyeditor Ruth Strother for smoothing out the rough spots and for all the great suggestions, and to the design team for bringing the images and text together so beautifully. Thanks to Jill Browning for her work on marketing and publicity.

To my family and friends, what can I say? Thank you for all the support, encouragement and interest in my quirky obsession with space exploration. To my wonderful, beautiful Mom, Artis and in memory of my Dad, Ken, thanks for always making me feel like I could do anything. To my siblings Alice, Mick and Linda: thanks for guiding your youngest sister along life's path and for your constant support. Thanks to my incredible mother-in-law Margaret and to all my wonderful brothers- and sisters-in-law, with special thanks to LaVon for the great pillows and the fun time we spent together during my interviews.

All my love forever to Andy, Nate and Nic and now Laurie and Jen: you've always been my inspiration and my most favorite people in the world to spend time with. "Like I've always told you, if you put your mind to it, you can accomplish anything!" To Collin, Connor, Landrie, Erin and all who may follow: you are the future; may you always be inspired to follow your dreams and go where they take you.

And to Rick, my constant guiding north star, my rock, and life's companion. I couldn't have written this book without your love and support. I love you always and forever.

ABOUT THE AUTHOR

Nancy Atkinson is the editor and writer for Universe Today, a popular space and astronomy news site, and is a NASA/JPL Solar System Ambassador. She has written thousands of articles about space, writing daily since 2004 about the latest news. She was the editor in chief for *Space Lifestyle Magazine* and also has had articles published on Wired.com, Space.com, NASA's *Astrobiology Magazine, Space Times* magazine and several newspapers in the Midwest. She has been involved with several space-related podcasts, including Astronomy Cast, 365 Days of Astronomy and was the host of the NASA Lunar Science Institute podcast. Nancy lives in Minnesota.

RECOMMENDED FOR FURTHER READING

NASA: www.nasa.gov
@NASA on Twitter and Instagram

ESA: www.esa.int
@ESA on Twitter. @europeanspaceagency on Instagram

JAXA: global.jaxa.jp/
@JAXA_en on Twitter

ISRO: www.isro.gov.in/
@ISRO on Twitter, @isro_india on Instagram

Roscosmos (Russian Space Agency): en.roscosmos.ru/
@roscosmos on Twitter, @roscosmosofficial on Instagram

Map and List of Space Agencies Around the World: pillownaut.com/spacemap/spacemap.html

Jet Propulsion Laboratory: www.jpl.nasa.gov
@NASAJPL on Twitter, @nasajpl on Instagram

New Horizons: pluto.jhuapl.edu/
www.nasa.gov/mission_pages/newhorizons/
@NASANewHorizons and @NewHorizons2015 on Twitter, @newhorizonsmission on Instagram
Books: *How I Killed Pluto and Why It Had It Coming* by Mike Brown, *The Case for Pluto: How a Little Planet Made a Big Difference* by Alan Boyle, *The Pluto Files: The Rise and Fall of America's Favorite Planet* by Neil deGrasse Tyson, *Chasing New Horizons: Inside Humankind's First Mission to Pluto* by Alan Stern and David Grinspoon (publish date Spring of 2017)

Curiosity Rover (Mars Science Laboratory): mars.nasa.gov/msl
@marscuriosity on Twitter & Instagram
Books: *Mars Rover Curiosity: An Inside Account from Curiosity's Chief Engineer* by Rob Manning and William L. Simon, *The Right Kind of Crazy: A True Story of Teamwork, Leadership, and High-Stakes Innovation* by Adam Steltzner and William Patrick, *Mars Up Close: Inside the Curiosity Mission* by Marc Kaufman, *Curiosity Rover: Design, Planning, and Field Geology on Mars*, by Emily Lakdawalla, (publish date Winter 2017.)

Hubble Space Telescope: hubblesite.org

www.spacetelescope.org

@HubbleTelescope on Twitter, @hubble_space on Instagram

Books: *Expanding Universe: Photographs from the Hubble Space Telescope* by Owen Edwards and Zoltan Levay, *The 4 Percent Universe: Dark Matter, Dark Energy, and the Race to Discover the Rest of Reality* by Richard Panek, *The Hubble Cosmos: 25 Years of New Vistas in Space* by David H. Devorkin and Robert W. Smith

Dawn Mission: http://dawn.jpl.nasa.gov/

Marc Rayman's Dawn Journal: http://dawn.jpl.nasa.gov/mission/journal.asp

@NASA_Dawn on Twitter

Kepler Mission and K2: kepler.nasa.gov/

www.nasa.gov/mission_pages/kepler/

@NASAKepler on Twitter

Book: *Five Billion Years of Solitude: The Search for Life Among the Stars* by Lee Billings,

Cassini Mission: saturn.jpl.nasa.gov/

www.nasa.gov/mission_pages/cassini

@CassiniSaturn on Twitter and Instagram

Book: *Titan Unveiled: Saturn's Mysterious Moon Explored* by Ralph Lorenz and Jacqueline Mitton

Solar Dynamics Observatory Mission:

sdo.gsfc.nasa.gov/

Helioviewer (and The Sun Now): sdo.gsfc.nasa.gov/data/

www.nasa.gov/mission_pages/sdo

SDO Science Blog: sdoisgo.blogspot.com/

@NASASunEarth on Twitter

Mars Reconnaissance Orbiter Mission:

mars.nasa.gov/mro/

HiRISE images: hirise.lpl.arizona.edu/

HiWish: uahirise.org/hiwish/

HiView: uahirise.org/hiview/

Beautiful Mars Project: uahirise.org/epo/

@HiRISE on Twitter

Lunar Reconnaissance Orbiter Mission:

lunar.gsfc.nasa.gov/

nasa.gov/mission_pages/LRO/

LROC: lroc.asu.edu/

@LRO_NASA on Twitter

Lunar and Planetary Laboratory: www.lpl.arizona.edu/

Books: *The Once and Future Moon*, by Paul Spudis, *A Man on the Moon* by Andrew Chaikin,

Juno Mission:

www.nasa.gov/mission_pages/juno/

JunoCam: missionjuno.swri.edu/junocam

@NASAJuno on Twitter

Satellite Servicing Capabilities Office: ssco.gsfc.nasa.gov/

Restore-L mission: ssco.gsfc.nasa.gov/restore-L.html

James Webb Space Telescope:

jwst.nasa.gov/jwst.stsci.edu/

Some of Nancy's Other Favorite Space Exploration Books:

The Overview Effect by Frank White, *Pale Blue Dot and Cosmos* by Carl Sagan, *Death From the Skies and Bad Astronomy* by Phil Plait, *Space: A History of Space Exploration in Photographs* by Andrew Chaikin, *Breaking the Chains of Gravity: The Story of Spaceflight Before NASA* by Amy Shira Teitel, *Bang! The Complete History of the Universe* by Chris Lintott, Brian May and Patrick Moore, *An Astronaut's Guide to Life on Earth: What Going to Space Taught Me About Ingenuity, Determination, and Being Prepared for Anything* by Chris Hadfield, *Lost Moon (Apollo 13)* by James Lovell and Jeffrey Kluger, *Rise of the Rocket Girls: The Women Who Propelled Us, from Missiles to the Moon to Mars* by Nathalia Holt.

INDEX